Sample Survey
Principles & Methods
Third Edition

Vic Barnett

Professor of Statistics, Nottingham Trent University, UK

A member of the Hodder Headline Group

LONDON

Distributed in the United States of America by
Oxford University Press Inc., New York

First published in Great Britain in 1974 as *Elements of Sampling Theory*
Second edition, *Sample Survey Principles and Methods*, published in 1991
This edition published in 2002 by Arnold, a member of the Hodder Headline Group,
338 Euston Road, London NW1 3BH

http://www.arnoldpublishers.com

Distributed in the United States of America by
Oxford University Press Inc.,
198 Madison Avenue, New York, NY10016

The advice and information in this book are believed to be true and
accurate at the date of going to press, but neither the author[s] nor the publisher
can accept any legal responsibility or liability for any errors or omissions.

British Library Cataloguing in Publication Data
A catalogue record for this book is available from the British Library

Library of Congress Cataloging-in-Publication Data
A catalog record for this book is available from the Library of Congress

ISBN 0 340 76398 1 (pb)

1 2 3 4 5 6 7 8 9 10

Production Editor: Rada Radojicic
Production Controller: Martin Kerans
Cover Design: Terry Griffiths

Typeset in 10/12 pt Times New Roman by Charon Tec Pvt. Ltd., Chennai, India
Printed and bound in Malta

What do you think about this book? Or any other Arnold title?
Please send your comments to feedback.arnold@hodder.co.uk

Sample Survey

Contents

Preface to the first edition

With the continual expansion of statistical theory, methodology and fields of application, it is inevitable that programmes of instruction in statistics cannot hope to present all aspects of the subject in full detail. A judicious choice of material is necessary. Rather than omit specific areas of study it is preferable, and indeed common, to cover some topics by means of short courses designed to provide a brief introduction to their methods and applications. Such courses present the student with a general survey of the topic and provide a springboard for more detailed individual enquiry, or for more specialised formal study at a later stage.

The collection and processing of data from *finite populations* is an important statistical topic from the practical and utilitarian standpoint, and can be a complex field of study in terms of statistical theory and methodology. Modern society abounds with published and broadcast reports of sample surveys which aim to describe the world we live in. In such surveys, samples are drawn from finite populations and are used to reflect on the population they claim to represent, or indeed are even extended in their claimed import to wider situations. Any student of statistics should have some acquaintance with the principles and methods that are used, and should be aware of the pitfalls of survey sampling. But unless he is specialising in this aspect of statistics it is unreasonable to expect his training to include a comprehensive treatment. Most likely this will be one of the subjects covered by the type of short 'special topics' course referred to above.

The study of *sample survey methods* contains many aspects. It involves setting up appropriate statistical principles, and constructing suitable statistical methods for collecting and analysing data from finite populations. But the implementation of such methods is bound up with sociological, psychological (and other) considerations. To know that a particular method of sampling and estimation is statistically desirable does not necessarily mean that it can be easily applied. Statistical propriety is essential. But such questions as practical access to the population of interest, the social acceptability of an enquiry, personal bias in the response to questionnaires depending on the formulation of the questions, all impose difficult non-statistical considerations. All these various aspects have been widely discussed individually in detailed texts. But what appears to be lacking is a concise modern treatment of the subject at an intermediate level. This monograph is designed to form the basis of a short course of instruction, perhaps constituting about 15 lectures. Inevitably it cannot fully scan the field, and it is directed principally to the study of the *statistical* aspects, whilst keeping in mind the problems involved in their application.

The book is based on short lecture courses given, in the Universities of Birmingham, Western Australia and Newcastle upon Tyne, predominantly to senior undergraduate and postgraduate students in statistics, but also on an interdisciplinary basis. It discusses the principles of different methods of probability sampling from finite populations in relation to their relative ease and efficiency for estimating properties of the population of interest. Chapter 1 considers some non-statistical aspects of survey sampling, and sets up the model for probability sampling. Chapter 2 is concerned with simple random sampling as a basis for estimating population means, totals and proportions. In Chapter 3 we discuss ratio and regression estimators which can exploit auxiliary information on additional variables in the population. Chapters 4 and 5 consider situations where further structure exists in the population and simple stratification or clustering methods may be appropriate. Some more complicated probability sampling schemes are described briefly in Chapter 6.

The emphasis is methodological; properties of different sampling schemes and estimators are discussed qualitatively as well as being formally justified. The treatment is at an intermediate level with mathematical proofs being heuristic rather than fully rigorous. A knowledge of elementary probability theory and statistical methods is assumed, such as would be obtained from an introductory course in Statistics. One unifying feature of the book is the empirical study of an actual simple finite population. Throughout the book the application and relative merits of different methods of estimation are demonstrated experimentally by constructing frequency distributions of estimates from this population. This augments and illustrates the theoretical discussion of the various techniques.

It is hoped that, as well as providing a basis for a short course for statistics students, the book may also serve as an introduction to the statistical methods of sampling for those involved in such work at a practical level in various fields of application, including business administration, medicine, psychology and sociology.

I am indebted to the Literary Executor of the late Sir Ronald A Fisher, F.R.S., and to Dr Frank Yates, F.R.S., and to Longman Group Ltd, London, for permission to reprint Table 1 from their book *Statistical Tables for Biological, Agricultural and Medical Research.*

It is a pleasure to acknowledge the help of friends and colleagues. I am grateful to David Brook and Betty Gittus for their useful comments on certain sections of the material, to Shiela Boyd for computer calculations, to Ray White for the two cartoons in Chapter 1, and to Shirley Daglish for her careful preparation of the typescript.

Vic Barnett
1973

Preface to second edition

The first edition of this book (under the slightly different title: *Elements of Sampling Theory*) has consistently served its intended purpose as a 'special-topic' or 'short course' text for the last fifteen years. Whilst its contents are as relevant as ever, there have inevitably been wide-ranging developments in sample survey principles and methods. It is essential, even in a necessarily selective treatment, to take such developments into account in offering a digestible but representative treatment of the field.

This new edition (entitled more appropriately *Sample Survey Principles and Methods*) adopts a broadly similar overall structure to the original one, but expands the coverage on several fronts to reflect present emphases and some new or extended methods. The major innovations include

- much greater attention to non-statistical, organisational, problems such as pre-survey sampling, sources of errors, obtaining the chosen sample, non-response, question formulation, and so on
- change of notation to more standard form
- extended coverage of different sampling schemes, in particular in the areas of ratio-estimation methods and multi-stage procedures
- wider study of more sophisticated probabilistic sampling schemes (such as sampling with probability proportional to size), and inclusion of some additional topics
- reinforcement of practical understanding by means of more substantial illustrations involving attitudes and opinions as well as objective and quantitative measures.

To ease the transition to the new edition, the overall structure of chapters and arrangement of topics has been retained as far as possible. The new material and emphases have been introduced largely by modifying or expanding the chapters of the first edition. The exceptions to this are the insertion of a new chapter on the practical considerations of *carrying out a sample survey* after the treatment in Chapter 2 of simple random sampling, and the dispersion of material on non-epsem schemes (such as pps sampling) throughout the book as appropriate rather than gathering it together as in the original concluding Chapter 6.

The broader canvas does, of course, provide material for larger courses of study (20–30 lectures, say) with no expansion of prerequisite demand. On the other hand, appropriate selection of topics still enables well-balanced shorter courses to be covered.

It is hoped that the updating and expansion of the contents of this new edition will make it attractive to student and teacher alike.

It is a pleasure to welcome a few more cartoons by Ray White in this new edition.

Vic Barnett
1991

Preface to the third edition

Sample survey principles and methods continue to evolve in response to new demands for wider areas of application and to advances in underlying research. This new edition aims to reflect such changes. There is a re-emphasis of various applications areas to reflect ever-growing concern for environmental and ecological issues and for financial and medical investigations, whilst continuing to stress the relevance of the methods in biological and social fields. Extra attention is paid to many topic areas and themes: with additional commentary on problems of non-response, randomised response, two-phase and two-stage sampling, panel surveys, focus groups, missing data and imputation, telephone and other electronic means of communication and access including e-mail and the internet, and many other topics.

The technical coverage extends earlier work on complex surveys including variance estimation. New chapters introduce modern approaches to sampling methods for rare and sensitive events and for natural phenomena with particular relevance to biological, environmental and social issues, including coverage of capture–recapture methods, composite sampling, ranked set sampling, transect sampling and weighted distribution methods.

Examples on all topics are given in the text and at the ends of chapter and numerical answers are provided in all cases. Chapter summaries and an expanded *Bibliography and references* (including early influential work as well as latest developments) improve the flow and extend the range of the methods. The material of the book is divided into four sections:

- Basic concepts
- Simple random sampling
- Practical aspects of carrying out a survey
- Environmental sampling

The aim of the new edition remains that of providing a digestible review of sample survey principles and methods at an intermediate level of treatment to serve students and practitioners. I am grateful to Marion Bown for computational assistance.

Vic Barnett
2002

Part 1
Basic concepts

1
Introduction

We start this first chapter by introducing some basic definitions. Our subject matter is *sample surveys* (e.g. opinion polls) from finite populations with the aim of learning more about the world we live in.

We will consider

- the nature of finite populations, surveys and opinion polls (1.1)
- specific examples of such means of enquiry (1.2)
- some basic concepts underlying sample surveys (1.3)
- why we should take samples (1.4) and how (1.5), and the central concept of *probability sampling* (1.6)
- a unifying practical example to guide us throughout the book (1.7)
- relevant references (1.8) and introductory exercises (1.9).

This age of the Internet and the World Wide Web is preoccupied with information. We seek ready access to data on all facets of our personal, social, environmental and professional lives. We need to express in quantitative terms all aspects of our lives, from cooking meals to political ideals. The communications media, including advertising, the Internet, television, radio, and the newspapers, maintain a constant flow of such information. No argument is complete without 'the figures to back it up'. Thus we may read (or hear) that:

> '*house prices in the UK increased by 14% over the last 12 months*',
> '*200 million viewers throughout the world watched the World Cup Final on television*',
> '*4 households out of 10 now have a home computer*',
> '*in 2000 almost 20% of workers in New York spent more than 2 hours of the working day travelling to and from work*.

The presentation of such figures is designed to keep us informed of the situation in the world around us; it is often used to support some proposal or criticism, or at least to place a discussion 'in the proper perspective'. Figures on drinking, or smoking, habits may be presented to encourage changes in the traffic laws, or as a partial explanation of variations in health in different sectors of the community. Results of opinion polls may be advanced as predictions of the outcome of a forthcoming election or to illustrate the need to change laws in relation to our social or physical environment.

Undoubtedly individuals are better informed than ever before, in the sense of being more exposed to quantitative descriptions of the world in which they live. This is a good thing, but it places serious demands on both the recipient and exponent of numerical information. On the one hand the 'man or woman in the street' needs to be able to understand and interpret the information that is presented. Some rudimentary knowledge of statistics is obviously desirable (and becoming more common). The ease with which data can be misrepresented or misinterpreted makes one sympathise with the old cry of 'lies, damn lies and statistics'—the prescription *cave emptor* has some justice! But there is a corresponding responsibility on those who present statistical data: to do so fairly and objectively with no intent to deceive, and to provide sufficient detail on the source, scope, and method of collection of the data for proper interpretation or further analysis.

The respective demands are not always met. In spite of an improvement in numeracy, the attitude to statistical data can still be one of bemusement or suspicion rather than of understanding or enlightenment, even though increasing awareness in the media and wider teaching of statistics in the schools are having their effect.

Then again, the presentation of data is not always flawless. Imprecise or incomplete statements, invalid inferences, graphs with distorted or unspecified scales (or relating to ill-defined concepts with psuedo-scientific names), and pictorial diagrams with psychological impact different from their factual basis all confuse the recipient. Whether such devices are deliberate, or merely arise through lack of statistical expertise, they can certainly serve vested interests to advantage. Standards continue to improve in these matters and perhaps the worst excesses are no longer to be found. The dangers are illustrated in an entertaining way by Huff and Geis (1993) and Moore (2001).

Such concerns reinforce the need for greater care in explaining statistical information, based in turn on sound knowledge and application of appropriate methods of collection, presentation, and interpretation of statistical data.

The aim of this book is to provide such knowledge and skill in a particular context: that of data arising from *sample surveys* or *opinion polls*. We will explore wide-ranging principles and methods with applications relating to all aspects of enquiry or interest. In the process we will assume however, that the reader has a basic knowledge of ideas and methods of probability and statistics such as that presented, for example, in the elementary or introductory texts on statistics by Freund (1992), Huntsberger and Billingsley (1987), Moore and Mc Cabe (1999), Mood, Graybill and Boes (1974) or Wetherill (1972).

1.1   Finite populations, sample surveys and opinion polls

The examples at the outset of the previous section illustrate a rather special type of situation and their study requires a rather special expertise. What characterises them is that they all relate to *finite populations* containing a *limited* and clearly defined set of *individuals* or individual units: the prices of all houses sold in the UK over the last 12 months, those people throughout the world with access to the particular television presentation, all people who work in New York, and so on. The population could be quite small (the 30 children in a school class or the 150 members of a golf club) or very large (perhaps many million) *but it is finite*.

Whether large or small, however, the aim is to say something about these finite populations by collecting and analysing information relating (in the main) to only a *part* of that population—what we call a *sample* from the population. This is obtained by *surveying* the population, and the study of how we should reasonably carry out such **sample surveys** is the special topic that we shall consider in this book.

Such sample surveys might cover any topic relating to the characteristics of a finite population, but we can broadly categorise the topic areas. Most surveys aim to describe human populations and their environment and fall into one of the following groups:

- demographic features of the population (e.g. family sizes)
- the economic structure of the society (e.g. industrial activity)
- patterns of life-style (e.g. travel habits)
- the social or physical environment (e.g. income or pollution levels)
- views and opinions (e.g. attitudes to central government).

Some basic distinctions must be drawn at the outset. In principle we could inspect *every member* of the finite population: we could strike *every* match in a box of matches (but note the effect of so destroying the population, a point we must return to) or enquire of every member of the golf club. Such a procedure is called a **census**: it is essentially a sample survey with 100% coverage! But typically we will need to be concerned with much lower levels of coverage—perhaps as low as 1% or 5% on occasions—since populations of interest can be, and often are, very large. Population sizes in areas where sample surveys are taken can range from a few hundreds or thousands to many millions.

Often the information we gather on individuals is quantitative and factual, perhaps describing some social or economic characteristics. Individuals may be people or may be light bulbs or farms, etc. When surveying human populations, information may include personal views or preferences. In this latter case the survey is commonly known as an **opinion poll**.

If we are concerned with the qualities of products or the attitudes of consumers then sample surveys can be vital, and survey sampling thus plays an important role in this field of **market research**.

The principles and methods of collecting and analysing data from *finite* populations is a branch of statistics known as **sample survey methods**; their formal basis is termed **sampling theory**. Whether the practitioner is an expert in a particular subject area (e.g. agriculture, industry, medicine or the environment) or a professional statistician, it is necessary to understand the basic principles and methods that underly the efficient study of finite populations.

Methods for such study differ in an essential way from most statistical work. Usually it is assumed that data arise as independent observations from an *infinite* population according to some probability model. In survey sampling there is a fixed, determined, *finite* set of individuals to be observed. Probabilistic considerations enter only if we impose them on our method of sampling.

Consider an example of this distinction. We are interested in the lifetimes of light bulbs of a particular type produced by a specific company. If we observe bulbs coming off the production line, standard statistical methods would lead us to propose some probability model for the lifetimes of the bulbs, such as an exponential distribution. Successive light bulbs would then be assumed to have lifetimes in the form of random observations from this distribution.

On the other hand we might have a carton of 100 bulbs. These bulbs have *specific determined lifetimes*, albeit unknown. In examining this *finite population* of 100 bulbs no *probabilistic* consideration need be involved. But we might impose one, by picking a bulb at random, for example. This is quite different from the structure assumed in the standard life-time distribution approach above. Note another complication. If we pick a bulb at random from the finite population and test it by using it until it fails, we have then changed the very population we are seeking to study (see *destructive testing* in §1.4 below).

In these pages we shall review the elements of such sampling theory, at a level which will be appropriate to those with some knowledge of the basic statistical ideas of probability distributions, estimation, and hypothesis testing.

1.2 Examples of sample surveys

We start by examining briefly some examples of sample surveys from different subject areas, illustrating the range of considerations and difficulties which may arise.

Agriculture

(a) The level of food (for example, fruit) prices may be a cause of some public concern. To study the current situation it is clearly not feasible to determine the prices charged at any particular time for every item of fruit sold by every supermarket or greengrocer or at every wholesale market in the geographic area of interest. But some indication of price levels and how they vary could be obtained by selecting a few types of fruit and enquiring about prices on a selective basis from different supermarkets, greengrocers or markets. But how should we effect this choice; what practical difficulties will we encounter; what will our survey tell us of the overall situation?

(b) A county council is required to submit an annual statement of the total wheat yield in its area. With great effort it might attempt a complete enumeration by contacting every farm. But there is no guarantee that it will receive a full response, or correct information, in each case. Just what is meant by a 'farm' anyway; is this a suitable unit for enquiry? On cost considerations it might again make sense to *sample* on some appropriate basis. But the sample will need to be adequate to meet certain requirements of *accuracy* and *validity*—it must reasonably 'represent' the population.

Education

Suppose an enquiry is to be conducted into the attitudes of schoolchildren to the subjects they are studying. Questions of interest might include: 'are you satisfied with your choice of subjects?', 'do you find some subjects more interesting than others?', 'what was the basis of your choice?', 'were you offered enough choice?', and so on. Cost and convenience again suggest a sample survey rather than a total national enquiry and existing census data will not contain the information we need. Such an opinion poll might be carried out for all schoolchildren in a particular metropolitan region and we might hope that its results would have a wider (possibly national) relevance. Attitudes may be expected to vary with the child's age, intellectual level, chosen subject combinations, and many other factors. An assortment of measures are simultaneously of interest. In the main, these are subjective rather than factual and will need to be elicited by individual enquiries in the form of **questionnaires** or **interviews**. The very way in which questions are asked can have a marked effect on the reaction of a child to the enquiry. Consider the three questions: 'Do you like History?', 'Do you dislike History?', 'Do you prefer History to Geography?': perhaps with a three-point attitude scale ('definitely', 'undecided', 'definitely not'). The responses to the different questions may show quite disparate attitudes to History as a subject of study. See Section 6.4 below for further discussion of this topic.

Environmental issues

Pollution is a major problem of our age, whether due to water impurities, leakage of radioactive substances or from our day-to-day exposure to heavy road traffic.

(a) In several metropolitan areas in Northern Italy, a large-scale survey showed that exhaust fumes from heavy road traffic may have adverse effects on the respiratory health of schoolchildren (Ciccone *et al.*, 1998). The children were seen to have increasing incidence of infections of the lower respiratory tract and of wheezing and bronchitis with increase in exposure to vehicular exhaust fumes. A survey based on a *questionnaire* given to a sample of about 40 000 children, comprising more than 25% of the reference population, produced a response rate of more than 90%. The results were *stratified* by level of urbanisation.

(b) Regular sampling of bathing water in recreational areas is carried out to check if water quality achieves required standards. As an example, weekly readings of mean pollution levels (e.g. of *e-coli*) might be taken in the form of a random sample of, say, five of the many monitoring sites around a lake. The sample mean level gives an estimate of the population mean level for the lake as a whole, but it is likely to be subject to high sampling variance and thus not readily suitable for checking if the standard is being maintained. Greater sample size would give greater precision but at greater cost.

More structured sampling methods which sacrifice randomness for informed choice of sampling sites can produce major gains in precision at no increase in cost.

Ranked set sampling is one such method. We choose at random five possible sites and seek (by guidance from local experts or by observations of other more cheaply measurable indicators) that one which *would give the smallest pollution level*. We measure the pollution at the chosen site. We do the same to identify and measure (from another five possible sites) the *second highest pollution level*, and so on up to the (*fifth*) *highest level*. So we again have a sample of size 5 for essentially the same effort as our random sample (assuming that costs of taking the sample and measuring its pollution level are much higher than identifying the sites for sampling).

Such ranked set sampling can give vast efficiency gains: e.g. with sampling variance just one third of that for random sampling.

Finance

Banking, insurance, accountancy, auditing and other financial specialisms are becoming more committed to mathematical and statistical methods, including those derived from quite complex stochastic models. Sample surveys are also widely encountered—not only on the marketing side but for financial processes in general.

In accountancy and auditing, the role of statistics is well established in principle. See, for example Arens and Loebbecke (1981), Smith (1976), PNMD (1989) and for a Bayesian approach to auditing, Steele (1992). In auditing, finite population sampling arises in the choice of transactions to be examined to augment other procedures of probity evaluation. A recent example of the use of complex sample surveys arose in a situation where it was necessary to combine information from the usual modest-sized externally generated audit sample with that from a much larger random sample obtained in an internal audit. See Barnett, Haworth and Smith (2001). Thus, suppose we are auditing a large account (population of transactions)—perhaps of 100 000 transactions or even more. An internal simple random sample of say 10 000 transactions has already been selected and each of the selected transactions has been assessed as 'correct' or 'in error'. It is possible in this internal audit that some of the assessments are *inaccurate* ('correct' transactions are in fact 'in error' and vice versa). Auditors then select a smaller simple random sample of say 1000 from the internal audit sample and *precise* assessments are made of whether these sampled transactions are correct or in error (and if so to what extent).

This is not an uncommon situation in practice but usually the auditors use only their own smaller sample of precise assessments to assess the audited account in regard to *error rate* and *total financial error*.

From a statistical viewpoint it is reasonable to examine what the (somewhat flawed) larger internal audit sample can add to our inferences about these two pivotal measures. Barnett, Haworth and Smith (2001) devise statistical methods for such an analysis of the two-stage sample survey data obtained in such a situation. They confirm that the internal audit can indeed add substantially to the inferential content. In fact, its evidence can sometimes swamp that of the external audit even when the attributions of 'correct' and 'in error' transactions in the internal audit are of significant order—say 10% inaccuracy in one or other (or both) respects.

Industry

(a) Market research is an important tool in the design of advertising campaigns, in the choice of types of product to be offered for sale, and in their manner of presentation. Public attitudes to products are commonly sought by means of sample surveys (*opinion polls* or attitudinal enquiries).

Suppose the topic of interest is reaction to the design and payment schemes of different types of mobile phones. We might expect that the different designs and payment schemes will appeal to different groups: to office workers, teachers, young professional people, retired persons and so on. It is important that our survey both reflects the views of the different groups and provides a representative coverage of the groups. Some people may resent being approached by interviewers in the street with such an enquiry and refuse to co-operate. *Quota sampling* is a technique commonly employed to cope with such problems of representativity and non-response by determining how many responses we want in each group and instructing interviewers to fill such 'quotas'. We will need to investigate the characteristics of such an approach.

(b) Monitoring the quality of manufactured products is a vast problem for industry. Except in the case of complex or expensive items full inspection cannot be justified. Investigation must inevitably be done on a sample basis, and reliable *sampling inspection* or monitoring schemes will be needed. In the past, *quality control methods* sampled and tested items on a random basis—from batches, or as produced on a production line. Such an approach is still used but is being replaced by a more holistic system of quality management in which the complete production process is permeated with statistical quality checks (*statistical process control*) to seek to ensure that the final product is up to specification. Parts of this more complex approach will need to be survey-based.

Medicine

(a) Great efforts are constantly being made to improve medical services in relation to their organisation, the standards of care, and the methods of treatment. Information on prevailing experiences and attitudes of patients and administrators is vital. Sample surveys are widely employed. This is another area where responses are likely to be seriously affected by human factors—psychological attitudes to doctors or health workers (veneration or impatience) and lack of full knowledge, or misunderstanding, of what is going on can confuse issues and lead to wrong answers. In a personal-interview enquiry about cervical cancer a woman was asked, 'Is your husband circumcised?' She replied, 'Yes doctor, *very*.'

(b) Surveys also play a major role in epidemiological studies of the state of health of whole communities (towns or counties or even countries). They might relate to disease incidence, treatment methods, morbidity or mortality. Surveys can link health and environmental issues as in studies of *social noise incidence* and its effects on hearing, particularly in young people. Noise is defined as 'unwanted sound'. The sound from clubs, pubs, cinemas and personal music players is 'wanted' by those who seek it (the club-goers) and very much 'unwanted' by those 'caught in the wash' (those living next to the club or pub). Surveys can tell us what is going on: what are current levels of 'social noise', are they worse than they were, and so on. They can also examine whether the hearing of young people is deteriorating due to extensive exposure to current levels. In recent surveys in the UK, noise levels experienced by 18–25 year olds were seen to be noticeably higher than 10 years ago. Detailed medical studies of surveyed individuals showed inadequacies in current standards and that increasing exposure to harmful sound levels (>97 dBA NIL), whilst not manifest in deterioration in threshold levels for hearing, is nonetheless leading to increases in incidence of tinnatus (particularly for males where the incidence rate is about 20%); (see Davis *et al.*, 1998).

Public utilities

Suppose a major public body, say a Water Authority, wants to estimate its likely expenditure over the next 10 years on repair and renewal to ensure maintenance of high-quality service on water supply, sewage disposal, etc. This will depend on the present nature and conditions of its stock of pipes, sewers, plant, and so on. Full inspection would be impossible: they cannot dig up every pipe and sewer! A sample survey, augmented by past records on expenditure, will be needed. But how are we even to define the *units* to be sampled (streets, local authority divisions, towns, etc.)? And for any defined and chosen unit, what do we *measure* and how do we relate the measurements to likely *future* expenditure? Imagine the cost of even a single observation: for examination, excavation, assessment of condition, and evaluation of investment needs for the next 10 years. It is hard to think of a more extreme example of the need to carefully design the survey to ensure a minimum possible sample size to meet required accuracy standards in the estimation of matters of the interest. Similar constraints operate across other areas of the utilities (and transport) sector: consider telephones, electricity, gas, trains, etc.

Social affairs

(a) We might wish to study the attitudes of 18 year olds in the UK to their newly acquired 'adult status' which includes the right to vote. A survey is to be conducted of this age group in a city area, relating to those whose permanent address is in the city.

Consider some of the difficulties in obtaining a sample for this purpose. Recognising that again we will need to sample the population rather than seek complete information, how are we to identify the members of the population we wish to sample? There is unlikely to be any complete official list of 18 year olds. The Electoral Register will not contain *all* current 18 year olds, in view of its method of notification being tied to some particular date. (In any case, it would be most inefficient to scan such a list in search of a sample of a mere minority of its members.) To stand on a street corner and interview people can quite clearly lead to an imbalanced or unrepresentative sample. *Non-response* apart, we will undoubtedly miss whole sections of our population—those away from the city street for various reasons, either on a temporary or semi-permanent basis (in hospitals, prisons, on holiday or working elsewhere temporarily). Injudicious timing of the enquiry can also lead to imbalance. During the week-day some will be at work, others at school, some in hospital or in prison, others unemployed—we may obtain many of the latter by street-corner interviews! At the weekend, or in the evenings, other sources of imbalance arise.

Clearly we would wish our sample to be representative of the different sections of the population. Direct approach to schools, hospitals, prisons, employment agencies, and so on, may be a more reliable and fruitful method of enquiry. But other difficulties now arise, principally concerned with obtaining access to the different sources.

(b) Family income and expenditure surveys are an important sociological guide. Consider some of the practical difficulties that might arise in studying family income in some geographical region. One fundamental matter which generates a deal of discussion is what is meant by a 'family'. Is it the same as a 'household'? How do we deal with multiple occupancy in houses, flats, or institutions? What items are included as 'income'? In short, just how do we define our *population* and the *measure* we wish to study? Then again personal resistance to such enquiries, and to the method of implementing a survey, can easily lead to imbalanced or inaccurate results. Door-to-door interviewing during the working day presents an obvious example—results are likely to lead to gross under-assessment. Again, consider the effects of inviting voluntary response to an income survey (or to current political or environmental issues)!

(c) A local government authority may wish to examine how its population reacts to critical aspects of law enforcement and crime in its region. Even more problems now arise. Non-response can be high because of the *sensitivity* of the issues. The *complexity* of and multi-faceted nature of the survey may exacerbate this and cost-factors may rule out personal interviewing. A *postal survey* might seem on first thought to provide a cheaper approach but possibly lower response rates from postal enquiries and the need for follow-up reminders may outweigh this apparent advantage. *Question design* in all-important, to avoid ambiguities and allay fears. Retaining confidentiality of response presents a major problem. Responses may be related to political views or income levels but it may be inappropriate (or counter-productive) to enquire about these matters. Perhaps *proxy questions* (such as on newspaper readership or what television programmes are watched) can help to provide such information. An initial, small, *pilot survey* may help to resolve some of the uncertainties. Would *telephone enquiries* or *door-to-door enquiries* be prudent?

Sample survey methods are widely used at all levels—interest groups survey their members, companies their possible customers, providers their clients and, particularly, authorities (from regional to national government level) their constituents. Most nations have extensive statistical services (such as the Office for National Statistics in the UK, the Australian Bureau of Statistics, Statistics Canada, and so on) to provide regular and comprehensive survey coverage; they may be part of the government/civil service structure or arms-length service providers.

The above illustrations have been carefully chosen to highlight the vast range of practical (essentially non-statistical) considerations and difficulties in designing and implementing a sample survey. They will prove useful for later reference in more detailed study of such practical considerations as:

- definition of a relevant population and what to seek to measure
- method of sampling (use of published data, interviewing, postal or telephone enquiries, etc)
- non-response and response bias of selected population members
- pilot surveys to aid the design of the main survey
- effects of questionnaire design and of the wording of questions
- choice of interview technique
- cost considerations
- use of supplementary information
- conducting and analysing complex surveys

Many of these *practical* topics are more fully considered in their own right in Chapter 6 (others arise naturally at appropriate stages in our study of the *statistical* bases of survey sampling).

1.3   Some basic concepts and definitions

Whilst the above examples highlight the various practical difficulties which can arise, they also point the need for some care in the more fundamental task of defining basic concepts and principles. We must start with these basic concepts and definitions. There is some inconsistency in the way different words are used. The following definitions serve our purpose in this book, but minor differences may sometimes be noticed in other treatments.

Fundamental to our studies is the idea of a *finite population*. *Individuals* in the population (not necessarily human; they might be light bulbs, or farms) have certain measures of interest. For example, our concern may be for the life-times of the bulbs, or the annual wheat yields of the farms. We would like to know certain *characteristics* of the population with respect to some measure—such as the *average* life-time of bulbs in the carton, the *total* wheat yield in Ohio, or the *proportion* of families in S.E. England with incomes in excess of £50 000 during the previous year.

Occasionally we may be able to derive the exact value of such a characteristic by studying every individual in our population. More often, limited time, money, or access dictates that we should *estimate* the characteristic by studying some smaller group of individuals in the population (a *sample* of its members) and infer the value of the characteristic from the information provided by the sample and by any general knowledge we have about the population.

Let us now commence our more detailed study by identifying some of the common *basic concepts*.

Target population

This is the total finite population about which we require information: for example, all 18 year olds in the UK.

Study population

This is the basic finite set of individuals we intend to study: for example, all 18 year olds whose permanent address is in the metropolitan area of our enquiry; or all wheat producers in Ohio in 1999. This may be (as in the latter case) the same as the target population. Alternatively (as in the former case) it may be a *more limited, more accessible*, population whose properties we hope can be extrapolated to the larger target population.

Population characteristic

This is that aspect of the population we wish to measure: for example, the *proportion* of 18 year olds who claim that they will exercise their vote at the next election, the *total* wheat yield in Ohio in 1999. This expresses some *aggregate* feature of the population in relation to how it varies from one individual to another. Each individual contributes a component (a number or qualitative description) for some *measure* of interest (voting intention, or wheat yield). Since this can vary from one individual to another we term it the *variable* of interest. The population characteristic of interest will usually be a *total*, *mean* or *proportion* of this variable over the population.

Sampling units

Ambiguities can arise concerning how we define or obtain access to individuals in the study population. Consider some examples. To investigate wheat yield in Ohio, we could sample fields or farms in the region (having decided how 'field' and 'farm' are to be defined), or perhaps sample larger administrative areas. Thus the potential members of the sample, the *sampling units*, can have different forms. They may be fields, farms, or administrative areas. A choice must be made at the outset of the enquiry; it can affect the usefulness of different sampling methods.

As a further example, suppose we wish to conduct a survey of family expenditure in some city. Although the 'individuals' in our study population are 'families', some conventional definition of 'family' must be adopted before we can proceed. Even so, there is likely to be no easy means of identifying or sampling such 'family' units. It would be far easier to sample *addresses* and to seek information on families at the chosen addresses. So the addresses become the sampling units, even though the population of addresses is not of essential interest.

Then again, in a survey on smoking and bronchitis in elderly people, we might most easily obtain information by approaching a sample of medical practitioners and asking about elderly people who have consulted them over a relevant period. The medical practitioners constitute the *primary sampling units*; their elderly patients are *sub-units* (which may be included in full in the survey, or further sampled—see *Cluster sampling* in Chapter 5; also *multi-stage sampling*). Note also that we are not sampling *all* elderly people, but only those who have visited their doctors.

Sampling frame

Thus the source of our sample is inevitably the *set of sampling units*. This is called the *sampling frame*. Sometimes the sampling units may be the individual members of the study population. Often this is not so and the sampling frame is a coarser sub-division of the study population, with each sampling unit containing a distinct set of population members.

List

To use the sampling frame as the raw material from which to draw our sample, we must be able to identify the sampling units. Indeed the sampling frame is chosen with this in mind. At best an actual *list* of all sampling units may exist, such as, for example, the list of city addresses, or the list provided by medical records of all elderly patients visiting their doctors in a given area over a certain period. Such a list makes it particularly easy to choose the sample. But if no tangible list is available for consultation, we must at least set up a *conceptual* list. For example, in studying prices of mobile phones, we will not have a list of all mobile phone suppliers to peruse at leisure. Nonetheless our list needs to consist of 'all mobile phone suppliers', and we must design our sampling scheme in such a way that it generates data from this list even though it may not be physically before us. This can be a complex matter. In any region there will be high-street shops specialising in mobile phones, but there are also general electrical suppliers and department stores which sell mobile phones. Then there is the 'virtual' world of the Internet where phones may be purchased electronically through computer links. All these must (or might) need to be included. Another, indirect, route is through agencies (such as banks) which claim to comprehensively review *all* prices of mobile phones (say) to find the cheapest ones for their customers. But those for different models may arise from different suppliers and we are unlikely to be given access to the full database from which the selections are made or even know its extent (it might be just a few suppliers).

Such distinctions are fundamental to the implementation of sample surveys. Let us refine and illustrate some of them by reference to the examples of Section 1.2. The problems they present include the following (in a roughly hierarchical relationship one to another):

   (i) choice of sampling units where various alternatives exist
  (ii) discrepancy between the ideal of a target population and the reality of an accessible sampling frame
 (iii) incomplete or intangible listing of sampling units
 (iv) implementation of the sample survey. Its organisation and administration involves a complex array of problems of planning, costing, and instruction. Furthermore, we should note the following problems:
      (a) if different types of individual exist, our sample should reflect these in a balanced way; there may be different problems in sampling the different types of individual. In the survey of attitudes of 18 year olds, those at school, in hospitals, at work, or unemployed, all present distinct sampling problems with regard to access, cost, and accuracy!
      (b) non-response in a sample survey can contaminate the results of the survey, as can psychological attitudes or inadequate understanding on the part of respondents to interviews or questionnaires in opinion polls
  (v) computer-based enquiries are also possible. There may be an electronic 'list' from which we could choose a sample. We would of course have to ensure that any such list is a reasonable sampling frame for our target, or study, population. For example, members of a professional group may be on a computer-based directory (such as UK statisticians on Allstat). But could we really sample UK statisticians at random by using the Allstat list? Not necessarily, for many reasons. If the directory conducts its business by e-mail then it contains all the e-mail addresses of its members. But these addresses are unlikely to be (allowed to be) revealed to an enquirer wanting to conduct a sample survey. Again, the Allstat members are UK statisticians, but not all

UK statisticians are on Allstat (some find it generates too much e-mail traffic, others may not be allowed by their employers to register with Allstat).

So whilst potentially the modern computerised and Internet-based electronic environment offers many new prospects for accessing lists relevant to specific sample surveys, in practice the opportunities for such methods of collecting survey samples are limited. Computerisation does have other major, realisable, advantages: for designing the survey, choosing the sample, storing survey data, analysing the results, etc. But whilst the computer provides speed and efficiency, and allows large-scale surveys to be handled in a possibly highly cost-effective way, it does not substantially change the principles of sample survey methodology.

The resolution of all such difficulties must be sought at two levels.

The *practical* (basically non-statistical) problems such as the choice of sampling units, administration of the survey, proper design of questionnaires, or adequate instruction of interviewers, and the like, require experience in a variety of applied disciplines. Detailed knowledge of the special features of the actual area of application of the survey (agriculture, medicine, social affairs, the environment, etc.) must be combined with advice from the psychologist on questionnaire design or on psychological test procedures, from the sociologist (or other appropriate expert) on the availability of relevant lists and records as a basis for the choice of sampling frame, and perhaps from the computer specialist on the automatic processing of the resulting data. A great deal of organised study, which can take us some way in avoiding the pitfalls, has gone into such matters. But we must ultimately depend on the native good sense of the organisers of a survey in the way in which they exploit the local circumstances and learn from their own experiences. Preliminary *pilot studies* in advance of the main survey can be a valuable aid.

In contrast, problems which relate to such matters as the representativeness of a survey, its validity, the choice of appropriate sampling procedures, methods of estimation of population characteristics (and the properties of these estimators) and legitimate interpretation of the results, all depend vitally on a proper understanding and application of *statistical* ideas. A sound *statistical* basis in the design of a sample survey is vital; 'practical' difficulties of implementation can reduce its effectiveness and must therefore be resolved as far as possible.

On the other hand, even a survey which presents no such 'practical' problems cannot be fully exploited if its statistical basis is inadequate; it is virtually *useless* if a total disregard of statistical design considerations makes it impossible to interpret or measure the accuracy of the results. Study of the appropriate statistical theory and methods *must therefore be viewed as the basic theme of the book*, and it is on this which we must now embark.

1.4 Why do we sample?

We have seen how the object of our enquiries is a population consisting of a *finite* number of individuals, on each of which some *measure* is observable. We want to characterise the population by some aggregate expression of that measure—perhaps its *mean*, or *total* over the population. It is natural to ask 'why not observe every individual in the population and thereby obtain the *exact* answer?'

In some cases, where the population is small and easily accessible, this is obviously a sensible policy. If I want to determine how much loose change I have in my pocket, I am hardly inclined to take a sample of the coins and to try to *estimate* the total value! Alternatively, a full enumeration (a **census**) may take place for a very large population when there is substantial social (financial, etc.) importance to justify the vast expenditure. This arises, for example, in national *Censuses* of (a limited number of) facts about *every member* of the country's population. But such cases are rare. More commonly it really does make sense, for a variety of reasons which we will examine, to restrict our study of the population to sampling *some* of its members and to using the information gained in this way to *infer* the characteristics of the population as a whole. Let us consider some of the reasons for using a **sample survey** rather than a *census*.

Cost

There will be a limit on the resources, in terms of money, time or effort, that we can apply. This is the main obstacle to a complete enumeration of the population. There is the need also to counterbalance precision and expense. Cursory inspection of a large number of individuals (possibly even the whole population) may yield, in view of inaccuracies of measurement, far less precise information than that obtained from more careful inspection of some judiciously chosen smaller sample. The use of alternative methods of medical testing provides a good illustration of this effect!

Furthermore we shall see how, even within the limitations imposed by some budget, different methods of sampling can yield (*for the same size of sample*) estimates of dramatically different precision. Finally we shall see how the additional precision which arises from increasing the sample size becomes, typically, less and less valuable in relative cost terms.

Differential cost factors are also relevant. In sampling the views of 18 year olds we may have to conduct face-to-face interviews with those in some group (e.g. in hospital) but write letters to those in another group (e.g. temporarily out of the area). The unit costs of sampling in these two different 'strata' are likely to be markedly different and the very *sampling design* which we choose to employ must reflect this difference: perhaps we will have to take a relatively smaller sample of those in hospital than of those who are away from home, or we may have to sample the former group in 'clusters' (all those in particular hospitals) to control contact and travel costs.

Utility

In some instances our sampling units may be destroyed in the process of sampling. Here a complete study of the population is sterile even if we can afford it. There is no point in knowing all about the population if it no longer exists for the application of our knowledge. Thus a manufacturer of light bulbs, or matches, is not going to test the lifetime of each bulb, or strike each match, to demonstrate the quality of his product. After such *destructive testing* there would be nothing left to sell!

Accessibility

Frequently there is different ease of access to different sampling units as we have remarked above. Some may not be even observable at all. Again we may be *compelled* to accept only a sample from the population. For example, historical records may be incomplete—temperature or rainfall readings over some period of interest may have been recorded sporadically; medical records may not contain all events of relevance or interest; contemporary attitudes to some controversial issue may have been incompletely recorded and we cannot recreate the circumstances for fuller study.

1.5 How should we sample?

This is the obvious next question to ask. Its resolution will require a more formal description of the finite sampling problem, and of the aims and objectives of a sample survey. This will occupy our attention for much of the remainder of the book. But we can usefully proceed a little further on a purely intuitive basis. The general aim must be to draw a sample which is a 'fair representation' of the population, and which leads to estimates of population characteristics with as great a 'precision' or 'accuracy' as we can reasonably expect for the cost or effort we are able to expend.

Various pragmatic or intuitively appealing methods of sampling have been advanced, and are still applied. Such *ad hoc* methods include the following.

Accessibility or haphazard sampling

With the prime stimulus of administrative convenience, a sample is chosen with sole concern for its ease of access. *We take the most easily obtainable observations*. The pitfalls in terms of lack of 'representativeness' are obvious. Consider the following examples.

(i) To study the sizes of lumps of coal brought to the surface in pit trucks, a few lumps are removed from the top of each truck.
(ii) In an opinion poll on colour prejudice, a notice is displayed asking for volunteers to answer the questions.
(iii) In an investigation of working habits of married women a door-to-door enquiry is conducted in a middle-class suburban area on a weekday afternoon.
(iv) Readers of a magazine are invited to complete and return a questionnaire published in the magazine.

The inevitable shortcomings of such samples as guides to the population as a whole are obvious in these examples. In other situations they may be less obvious—but they may be no less serious!

Judgmental or purposive sampling

The attitude here is quite different. Recognising that the population may well contain different types of individual, with differing measures and ease of access, the

Mrs Smith, personally selected as a typical, ordinary housewife

experimenter exercises *deliberate subjective choice in drawing what he or she regards as a 'representative' sample*. The results of such a sampling procedure *can* be very good, if the experimenter's intuition or judgment is sound and it has to be recognised that some surveys may employ this principle to some extent. In particular, ranked set sampling (see Chapter 7) benefits from such a prospect but in a mathematically structured way.

Judgmental sampling aims at the elimination of *anticipated* sources of distortion; but there will always remain the risk of distortion due to personal prejudices, or to lack of knowledge of certain crucial features in the structure of the population. This latter factor is well illustrated by the presence of *unrecognised* correlations between the criteria of choice of the sample and the measure being studied. This arises in the classic example of sampling ash content in coal by taking some coal from the edge of *each* of a set of piles of coal in order to obtain a 'representative' sample of the different piles of coal. But the ash content varies with the size of the lumps of coal. So, whilst representing the different piles, the sampling procedure ignores the effect of size. We will tend to pick smaller lumps (at the edges of the piles) and the sample will be far from 'representative' in this other crucial respect.

Similar distortions can also arise in phone or postal enquiries. Even when we design a randomised approach, patterns of non-response can conspire to produce an unrepresentative set of responses.

Quota sampling

Often judgment and accessibility are combined. For example, in *quota sampling* (see Section 4.7) people may be interviewed in the street in an attempt to obtain a sample judged to be well representative of different ages, sex, occupations, and so on. This involves an element of accessibility: the 'most promising looking passers-by' are chosen to fill the quotas. This method is however, much more structured and sound than straight accessibility or judgemental sampling. A proper statistical design will have been used to determine what numbers are needed in each of the quotas and the aim will be to fill the quotas on a randomised basis. However, some subjectivity and arbitrariness of choice of their constituent members cannot be avoided even with the most careful instruction of interviewers. Quota sampling is the most widely used approach to opinion polls, market research studies, and so on, even by the major survey organisation such as National Opinion Polls, MORI and Gallup, in the UK, and by broadcasting audience measurement concerns. We will need to discuss quota sampling in more detail later.

The major criticism of accessibility and judgmental sampling is not that they may lead to unrepresentative samples, but that their results are unconvincing (and too easily dismissed if unpalatable) *because there is no yardstick against which to measure 'representativeness' or to assess the propriety or accuracy of estimators based on such a sampling principle.* Such a yardstick is vital!

For this reason we are compelled to introduce an element of 'randomness' into sampling procedures and to draw our samples according to some imposed probability mechanism. Such an approach is essential if we are to have a rationale for describing the representativity and precision of our survey and its resulting estimators. A variety of *probability sampling schemes* have been devised, evaluated, compared, and utilised, and we shall study these in some detail as the only sound basis for survey sampling.

To proceed with this we must begin to formalise our finite population model and our sampling objectives.

1.6   The central concept: probability sampling

Let us suppose that, in an effort to study some finite target population, we have resolved the matter of the choice of appropriate sampling units and of the sampling frame they comprise. Suppose that the sampling frame represents the accessible finite population, and that the sampling units are the individual members of that population. We shall refer merely to the 'population' and its 'members' or 'individuals'. Our interest centres on the values taken by some variable, Y, for the different members of the population, and on aggregate measures of this variable over the population. Thus if there are N members, we can represent the population by

$$Y_1, Y_2, \ldots, Y_N,$$

these being the values of Y taken by the different members.

We will be interested in **population characteristics** defined with reference to Y. Those most commonly studied are:

(i) The population *total*, $Y_T = \sum_{i=1}^{N} Y_i,$

(ii) the population *mean*, $\bar{Y} = \dfrac{1}{N}\sum_{i=1}^{N} Y_i = Y_T/N$,

and

(iii) the *proportion*, P, of members of the population which fall into some category of classification for the measure Y.

For example, in a social survey of car-driving habits in an adult population, P may be the proportion who drive more than 10 kilometres each day.

The aim of the sample survey will be to *estimate* one or more of these characteristics from the information contained in a sample of n ($\leq N$) members from the population. Suppose the values of Y for the sample are

$$y_1, y_2, \ldots, y_n$$

where each y_i is one of the values Y_j, of Y, in the population at large. Not all Y_j are necessarily different; the same is true of the y_i.

Strictly speaking, different y_i *might* arise from the *same* Y_j (if we are sampling 'with replacement' in the sense that a particular population member could be chosen more than once in the sample) but we shall assume (other than in parts of Chapters 2 and 5) that this is not so.

So, unless otherwise specifically stated, the sample is assumed to have been drawn '*without replacement*'—once an individual member of the population has been chosen it cannot be chosen again. This is the usual situation in survey sampling. (The parallel study of sampling with replacement uses traditional statistical ideas and does not involve many of those special considerations relating to sampling without replacement from a finite population. However, it will be informative later to illustrate some of the distinctions that arise in the two cases.)

Although Y may in practice be *multivariate* (consider the response of an individual to a large set of questions in a questionnaire), we shall concentrate almost entirely on situations where Y *is univariate*. Problems of simultaneous estimation of characteristics of the components of multivariate Y, or of measures of associations between these components, will not be directly considered. But we will later consider how *joint observation of two associated variables* may lead to more efficient estimation of the characteristics of one of them (by exploiting any relationship between then) and also how to estimate ratios of characteristics of the components of such bivariate Y. (See Chapter 3.)

Terminology

The ratio of sample size to population size

$$f = n/N$$

will be called the *sampling fraction*. This is a useful measure of sampling effort.

To estimate Y_T, $\bar{Y}$, or P we will need to calculate some summary measures of the sample. Thus to estimate $\bar{Y}$ it might seem appealing to use the *sample mean*

$$\bar{y} = \frac{1}{n}\sum_{i=1}^{n} y_i.$$

But how are we to assess the properties of such an *estimator*? One possibility is to enquire how values of $\bar{y}$ may vary in relation to $\bar{Y}$ from one occasion to another when we employ the current sampling procedure in the same problem. However, with accessibility or judgmental sampling we cannot answer such a question; the lack of any objective sampling principle amenable to repetition rules it out.

Consequently we are led to introduce some probability mechanism as a means of drawing samples, and must consider the idea of:

Probability sampling

In a general **probability sampling scheme**, we firstly specify the size, n, of sample to be drawn. We then consider (conceptually at least) *all possible samples of size n* that could be drawn from the population; $S_1, S_2, \ldots$. Thus, each S_i is a *distinct* sample of size n *which could be drawn from the whole population*. Note that 'distinctness' relates to population membership, not necessarily to the values taken by the measure Y.

The *probability sampling scheme* is defined by assigning a probability π_i to each S_i and a particular sample, S, can then be chosen in accord with this probability scheme.

A vast assortment of different probability sampling schemes are possible, corresponding to different probability distributions $\boldsymbol{\pi} = \{\pi_1, \pi_2, \ldots\}$ over the set of possible samples, $S_1, S_2, \ldots$. These range from simply defined and implemented schemes to highly sophisticated ones.

We shall be considering an assortment of the more straightforward schemes that are commonly used, and comparing them in terms of their costs and how well (in a sense to be defined) they enable us to estimate $\bar{Y}$, Y_T, and so on.

What would we want of an estimator? It should provide a *representative* and *accurate* estimate. With such a probability-based sampling principle we are now able to discuss the 'representativeness' of the sample (in terms of its method of generation) and the 'accuracy' of estimators, employing the usual statistical concepts.

Suppose that θ is some population characteristic (it may be Y_T) and that we choose to estimate it by some function, $\tilde{\theta}(S)$, of the sample, S_1. $\tilde{\theta}$ is called a *statistic* or *estimator*. We can discuss the properties of the sampling scheme and of the estimator in terms of the **sampling distribution** of $\tilde{\theta}$ induced by the probability distribution $\boldsymbol{\pi}$. Different values of $\tilde{\theta}$ will be encountered on different occasions, with probabilities determined by $\boldsymbol{\pi}$. This set of possible values and their associated probabilities is the *sampling distribution*.

Unbiasedness

One possible criterion on which the sampling scheme may be judged 'representative' is that $\tilde{\theta}$ should be **unbiased**. That is to say, on average $\tilde{\theta}$ should be θ.

So we want

$$E_\pi[\tilde{\theta}(S)] = \theta,$$

where E is the expectation operator taken over the sampling distribution of θ.

We shall give prime attention to unbiased estimators. Only occasionally are we prepared in sample survey work to 'trade bias for precision'.

Precision

Often the estimator $\tilde{\theta}$ has, at least in large samples, an approximately normal distribution. It is reasonable, therefore, to assess the 'accuracy' or 'precision' of an unbiased estimator by considering its **variance**,

$$\text{Var}[\tilde{\theta}(S)] = E_\pi\{[\tilde{\theta}(S) - \theta]^2\}.$$

The smaller this variance, the more 'precise' the estimator. If, for a given size of sample, one unbiased estimator has lower variance than another, we say it is *more efficient*. In this way we can compare estimators or different probability sampling schemes.

If $\tilde{\theta}$ is biased, $\text{Var}[\tilde{\theta}(S)]$ needs to be replaced by the *mean square error*,

$$\text{MSE}[\tilde{\theta}(S)] = E_\pi\{[\tilde{\theta}(S) - \theta]^2\}$$

—the smaller this quantity, the better (or more 'precise') the estimator. Very occasionally, a biased estimator with small MSE may be preferred to an unbiased one with larger variance, but as remarked above we shall concentrate attention on unbiased estimators in the main.

The broad aim of sampling theory is thus to devise sampling schemes which are economical and easy to operate, which yield unbiased estimators, and which minimise the effects of sampling variations.

The effects of sampling variations is reflected by $\text{Var}[\tilde{\theta}(S)]$ for an unbiased estimator. In general $\text{Var}[\tilde{\theta}(S)]$ decreases with increase in sample size, but the cost increases. We must effect a balance. We shall be comparing sampling schemes to determine which of them yields an unbiased estimator with smallest variance *for a given cost*, or *for a given sample size*. We must also tackle the inverse problem of choosing the sample size to yield prescribed *precision* (in terms of $\{\text{Var}[\tilde{\theta}(S)]\}^{-1}$).

Note that on the approach described, all probabilistic behaviour is defined in terms of the sampling scheme alone. Inferences about population characteristics are drawn on this basis and are termed **randomisation inference**. In parametric inference for infinite populations, it is usual to model a random variable, Y, of interest in terms of some family of distributions $\mathcal{P} = \{p_\theta(y) : \theta \in \Omega\}$ indexed by a parameter θ in a parameter space Ω which describes probabilistically how observations y of Y have arisen. Formal inference procedures, termed **model-based inference**, are based on the probability model.

It is possible to augment the probability sampling scheme in finite population sampling by model-based assumptions about the constituent members of the finite population. We will return to this in Section 2.14 below.

The ease of operation and administration of a sample survey is an important aspect of cost reduction. We shall sometimes find that sheer administrative convenience (rather than direct concern for precision) can override other factors in promoting a particular method of sampling. (See *cluster sampling*, *systematic sampling*, and *quota sampling.*)

1.7   A real-life study: the National Team

To illustrate the theoretical results obtained throughout the book, we shall simultaneously construct empirical sampling distributions from a fully specified finite population. By considering this actual population, we will be able to see just how anticipated theoretical properties of sampling schemes actually do arise in practice.

Our population is the *National Team* which is about to go off to the *Global Games.* The team is made up of 50 athletes whose skills cover track and field events. The numbers of men and women, and sprinters, long-distance runners, single event field athletes and multi-eventers (e.g. Heptathlon) are shown in Table 1.1.

Certain physical measures of the team are important for monitoring progress: in particular *weight.* We will consider, as we discuss different sampling methods, the effect of estimating average weights by these different methods. To do this we assume that there is a list of these basic measures for the whole team (i.e. for the complete population) in the form of the *genders*, *weights* and *heights* of all members. Some of the more sophisticated sampling methods (such as *ratio estimators*, *stratified estimators* or *cluster sampling estimators*) will take account of these characteristics and of the events for which the individual members are entered (*sprint*, *long-distance*, *field* or *multi-event*).

Table 1.2 shows the events and the physical measures for the population along with relevant averages, variances and correlation coefficients. Weights and heights will be denoted Y and X, and measures are in kg and cm, respectively.

Of course, our *Global Games* team example is not realistic since if we knew the values of the relevant variables for the complete population we would not need to *estimate* population characteristics from *samples*. However, it provides the opportunity, when examining different sampling schemes and their relative properties, to

Table 1.1 The team for the Global Games

Event	Male	Female	Totals
Sprint	8	6	14
Long-distance	10	6	16
Field	6	4	10
Multi-event	6	4	10
Totals	30	20	50

Table 1.2 Statistics for the Global Games team

Event	Male		Female	
	Weight	**Height**	**Weight**	**Height**
Sprinting	70	184		
	74	169	55	160
	73	183	54	170
	73	164	57	171
	78	196	58	155
	70	188	49	151
	72	173	57	159
	76	192		
Long-distance	73	189		
	57	152		
	74	171	64	162
	69	165	53	160
	72	159	56	166
	69	169	67	180
	81	195	63	175
	75	192	71	179
	58	179		
	69	164		
Field	82	175		
	89	198	59	159
	70	176	68	169
	78	187	68	182
	76	168	64	175
	76	192		
Multi-event	71	184		
	77	171	63	175
	67	168	65	151
	71	180	60	173
	72	188	56	159
	63	168		
Number	30		20	
Mean	72.50	177.97	60.35	166.55
Variance	41.22	148.31	34.56	92.79
Correlation	0.557		0.617	
Overall		*Weight*	*Height*	
Mean		67.64	173.40	
Variance		73.95	155.67	
Correlation		0.683		

observe in numerical terms (albeit for a small and artificial population) the types of distinction our theory tells us we should expect. We shall keep revisiting our *Global Games* team from this point of view as we explore the different sampling schemes and approaches.

Example 1.1

We can make our first use of the *Global Games* data by examining how the two intuitive sampling methods might work out.

No prescription can be given for how to draw an **accessible sample** or **judgmental sample**. The type of sample chosen will vary with the whim of the sampler. This is the disadvantage of such methods.

However, suppose it is the team coach who is monitoring weight and decides to take a sample of five athletes to estimate mean team weight. The female sprint team happen to be having a strategy chat outside his window: what better (more accessible) sample than this! So he obtains an estimate of $\overline{Y}$ in the form $\overline{y}_A = 281/5 = 56.2$ kg which is, of course, a very low value compared with the population mean of 67.64. But this is hardly surprising—all sample members are female and from the group (sprinters) which are likely to be most lithe.

But suppose he decided to exercise his judgment to avoid such bias or lack of representativity. So he deliberately picks three males and two females from different sections of the team. Suppose he obtains a judgmental sample which gives a sample mean of $\overline{y}_j = 403/6 = 67.17$ kg. This is much closer to the true value than the accessibility sample mean.

But we have no way of assessing the sampling procedure in terms of values of $\overline{y}_J$ which might arise. Presumably, if the chosen sample has the maximum appeal to the sampler, he will always choose this sample. Thus he *always* obtains 67.17 as his estimate. So the sampling procedure will lead to a good estimate if it leads to a good estimate, and vice versa! We can say no more regarding precision.

1.8 Selected bibliography and references

The treatment of sample survey theory and methods presented in this book is intended to be complete and self-contained in the respects outlined in the Preface. Nevertheless, the reader may wish to persue slightly wider or more detailed study of some of the topics. The following books and journal publications might prove helpful in this respect. They are appropriately categorised by emphasis or content.

General Introduction and Motivation. Kalton (1983); Kish (1995); Stuart (1984).

Statistical Theory and Methods. Cassel *et al.* (1977); Cochran (1977); Hansen, Hurwitz and Madow (1993); Hedayat and Sinha (1991); Levy and Lemeshow (1991); Kish (1995); Schaeffer, Mendenhall and Ott (1986). Additionally, references are given at appropriate points in the text to immediately relevant published material—we have already encountered some in this opening chapter. Such references include original or early work on many of the themes.

Practical Considerations (general). Biemer *et al.* (1991); Dillman (1999); Groves and Couper (1998); Groves *et al.* (1988, 2001); Hajek (1981); Jessen (1978); Groves (1989); Lessler and Kalsbeek (1992); Singh & Chaudhury (1986); Thompson (1992); Hansen, Hurwitz and Modow (1993); Hoinville, Jowell *et al.* (1978); Kalton (1983); Kasprzyk *et al.* (1989); Moser and Kalton (1971, 1999); Raj (1972); Yates (1981). For further details on references to specific topics (e.g. telephone surveys, panel surveys, response errors) see Chapter 6.

Complex Surveys. Brewer and Hanif (1983); Lehtonen and Pahkinen (1995); Skinner *et al.* (1989); Wolter (1985).
Accountancy. Smith (1976).
Agriculture. Sampford (1962).
Biology. Buckland *et al* (1993); Cormack *et al* (1979); Seber (1982, 1986).
Business and Industry. Deming (1990), Cox *et al.* (1995).
Environment. Barnett (2003); Gilbert (1987); Journel (1988); Stehman and Overton (1994); Webster and Oliver (2000).
Health. Levy and Lemeshow (1980); Korn and Graubard (1999).
Psychology. Nunnally (1967).
Sociology (and Politics). Hoinville, Jowell *et al.* (1978); Moser and Kalton (1971, 1999).

(See *Bibliography and References* for publication details).

1.9   Exercises

1.1   For the illustrative examples, *Medicine* (b) and *Social Affairs* (b) in Section 1.2 discuss the difficulties underlying the choice of an appropriate sampling frame, and the organisational problems likely to be encountered in collecting data from your preferred frames.

1.2   In a national enquiry into Health Service needs, it is necessary to study, through an appropriate sample survey, the pattern of general medical practice. We want to examine the 'hospital referral' pattern for general practitioners over a particular year: how many patients they send to hospital for inpatient or outpatient treatment; how many to consultants for specialist comment, or to psychiatrists; how many need the services of health visitors; and so on. Two alternative sampling schemes are available: complete enumeration for a small sample of practitioners, or a sample of individual patients throughout the country. Discuss the possible advantages and disadvantages of the two schemes.

1.3   We wish to examine the views of a large class of undergraduate university students on a lecture course they are taking on Sample Survey Methods. The aim is to use this feedback of information to redesign the course (if necessary) for next year, in as far as the method of presentation of the material effects its impact and understanding. Construct a short questionnaire for circulation to the members of the class for this purpose. Should the questionnaire be given to all class members, or to a sample? What should we do about non-response?

1.4   (*For class co-operation*). Choose an *accessible* and a *judgmental* sample of size 5 from the *Global Games* team for the purpose of estimating mean weight explaining how your samples might satisfy the *accessible*, and *judgmental* descriptions. If this example is being carried out in a teaching context, contribute your estimates to the results for the whole class and construct histograms in the two cases. Discuss what appear to be the factors which have influenced the choice of samples by the two methods.

1.5   Suppose it is necessary to draw a sample:

(i) of telephone subscribers whose telephone numbers refer to a particular exchange, from a directory which covers many exchanges in addition to the one of interest

(ii) of members of an e-mail bulletin board list

or

(iii) of a particular set of authors, from a composite hand-written list of all books published by those authors.

Discuss any difficulties which might arise in drawing reasonable samples in each case, and suggest methods of overcoming such difficulties. How would you try to contact the individuals in each case?

Part 2
Simple random sampling

2
Probability sampling and simple random sampling

We begin to develop the formal machinery of probability sampling and of estimators based on a probability sampling scheme by defining and exploring the most basic model of *simple random (sr) sampling*.

- Simple random sampling is defined (2.1) and seen to be equivalent to choosing population members at random without replacement one at a time until the required sample size is obtained.
- We see (2.2) that the sr mean $\bar{y}$ is unbiased for the population mean $\bar{Y}$ with variance $(1 - f)S^2/n$ where S^2 is the *population variance* and f is the *finite population correction*. We see that $\bar{y}$ is the *best linear unbiased estimator* of $\bar{Y}$.
- Random sampling with replacement produces different results from sr sampling (2.3).
- Estimation of the population variance S^2 is important (2.4) for assessing the precision of $\bar{y}$, for comparing it with other estimators and for determining how large a sample is needed to achieve prescribed precision.
- We see how to determine confidence intervals for $\bar{Y}$ (2.5) and how to make a reasoned choice of the sample size n (2.6).
- Systematic sampling (2.7) and non-epsem sampling (2.9) are important variations which we need to begin to explore.
- The results for $\bar{Y}$ can be readily extended to the estimation of the population total Y_T (2.8) and a population proportion P (2.10), including confidence intervals for P (2.11) and choice of sample size for prescribed precision (2.12).
- Some further matters concerned with estimating proportions are set out (2.13)
- We need to draw a fundamental distinction between *randomisation inference* and *model-based inference* for handling finite populations and to stress the importance of the *design* of the sample survey (2.14) for later relevance before considering some more *Exercises* for the reader (2.15).

The most basic form of probability sampling is *simple random sampling*. It is widely used in its own right and is easy to operate from the statistical viewpoint. It also serves as the basis for more complicated sampling schemes, such as *stratified simple random sampling*, and *cluster sampling*. The properties of estimators obtained from

simple random samples may be readily demonstrated, and this will be the principal
object of the present chapter.

2.1 The simple random sampling procedure

This operates in the following way. If the population is of size N, and we require a
simple random sample (srs) of size n, this sample is chosen *at random* from the $\binom{N}{n}$
distinct possible samples, in each of which no population member is included more
than once.

That is to say, each of the $\binom{N}{n}$ samples has the same probability $\binom{N}{n}^{-1}$ of being
chosen. Simple random sampling is an example of what is sometimes referred to as an
epsem: an **equal probability selection method**, where each population member has
the same probability of appearing in the sample. But more complicated sampling
schemes can also be *epsem's*. This is true, for example, in the cases, discussed later,
of *stratified simple random sampling* or *one-stage cluster sampling* employing the
probability-proportional-to-size (*pps*) principle.

Such a sample can be obtained sequentially: by drawing members from the population one at a time *without replacement*, so that at each stage every remaining member of the population has the same probability of being chosen. We will assume that
the term 'simple random sampling' implies sampling at random *without replacement*.

We see that this produces a *simple random sample* as follows. Suppose this sequential method of choice yields n (distinct) population members whose Y values are

$$y_1, y_2, \ldots, y_n$$

where y_i refers to the ith chosen member ($i = 1, 2, \ldots, n$).
    The probability of obtaining this *ordered* sequence is

$$\frac{1}{N} \cdot \frac{1}{N-1} \cdots \frac{1}{N-n+1} = \frac{(N-n)!}{N!}$$

But any reordering of $y_1, y_2, \ldots, y_n$ corresponds to the same choice of n distinct population members (that is, the same sample); there are $n!$ possible reorderings. Thus the
probability of obtaining any particular set of n distinct population members (irrespective of order) is just

$$\frac{n!(N-n)!}{N!} = \binom{N}{n}^{-1}.$$

There are $\binom{N}{n}$ such sets (or *samples*) which can arise, and these samples are therefore
generated with equal probabilities: that is, *by simple random sampling*.
    The choice of individual observations in the sample is achieved at each stage by an
appropriate random mechanism applied to the remaining members of the population;
for example, using a table of random digits such as that given as Table 1 in the
Appendix or the random number generating facility on any modern computer. Thus if
we want a srs of size n from a population of size N, we first choose a digit at random
in $(1, 2, \ldots, N)$ to identify the first sample member. We then choose a digit at random

in $(1, 2, ..., N - 1)$ to locate in the remaining population our next sample member, and so on. To choose a digit i at random in $(1, 2, ..., M)$ from a random uniform deviate X on $(0, 1)$ we would take an observation x and form $i = [Mx + 1]$ where [ ] means the integral part of. A corresponding principle applies if we have available (as in Table 1 of the Appendix) a set of random digits e.g. we can group them in sets of k where $10^{k-1} \leq N < 10^k$ and choose n consecutive *distinct* k-tuples each in the range 1 to N inclusive—thus identifying our srs of size n in a single operation. Alternatively, many statistical computer packages will have designed facilities for choosing a random set of n values from 1 to N.

Example 2.1

Choose a simple random sample of five weights from the *Global Games* team in the example of Section 1.7. Numbering the population members 1 to 50 down the columns of male and then female weights in Table 1.2, we find that the first five distinct pairs of numbers from 1 to 50 in Table 1 of the Appendix (reading along the rows) are 23, 32, 28, 24 and 33 yielding a simple random sample 76, 54, 76, 71 and 57 of weights.

In such a simple situation this is a reasonable task. But notice how for larger populations and samples the effort of choosing the sample from a table of random digits can be quite excessive. If N is not 10, 100, 1000, and so on, we must either disregard a possibly large portion of the table, or allow multiple reference from the table to individual population members. We must also keep an account of those members which have been chosen and ignore them on subsequent random appearance. These factors make it desirable to give careful thought to the actual mechanics of using a table of random numbers or a computer random number generator to choose the sample, in order to keep the effort as low as possible.

We shall consider the use of simple random samples for estimating the three population characteristics: the *population mean* $\bar{Y}$, the *population total* Y_T, and the *proportion P*, of Y-values in the population which satisfy some condition of interest. We shall need to discuss how any estimators behave in an aggregate sense: that is, in terms of their sampling distributions. By analogy with the study of traditional estimators of parameters in infinite population models, where the variance of the parent population is often a crucial measure, we need at the outset to define the *variance* of a finite population.

Variance

The variance of the finite population $Y_1, Y_2, ..., Y_N$ is

$$S^2 = \frac{1}{N-1} \sum_{i=1}^{N} (Y_i - \bar{Y})^2.$$

Notice that this is a *deterministic* measure, not a probabilistic average as it is in the case of random variable theory for probability models.

The divisor (here $N - 1$) is arbitrary. In other treatments of sampling theory sometimes $N - 1$ is used, elsewhere N. All we need is some convenient measure of the variability of the Y values in the population. S^2 is particularly convenient in that it leads to simpler algebraic expressions in the later discussion, and produces results more closely resembling corresponding ones in the infinite population context.

The concept of probability averaging only arises in relation to some prescribed probability sampling scheme. Thus for *simple random sampling* we have the concept of the *expected value* of y_i, the ith observation in the sample. That is

$$E[y_i] = \sum_{j=1}^{N} Y_j \Pr(y_i = Y_j) = \frac{1}{N} \sum_{j=1}^{N} Y_j = \bar{Y}.$$

The result that $\Pr(y_i = Y_j) = 1/N$ holds because the number of samples with $y_i = Y_j$ is $(N - 1)!/(N - n)!$, and each has probability $(N - n)!/N!$

We easily see that

$$E(y_i^2) = \frac{1}{N} \sum_{j=1}^{N} Y_j^2,$$

and

$$E(y_i y_j) = \frac{2}{N(N - 1)} \sum_{r<s} Y_r Y_s \qquad (i \neq j).$$

Hence the *variance* of y_i, and *covariance* of y_i and y_j, are

$$\begin{aligned}
\mathrm{Var}(y_i) &= E\left[\left(y_i - \bar{Y}\right)^2\right] \\
&= E(y_i^2) - \bar{Y}^2 \\
&= (N - 1)S^2/N
\end{aligned} \tag{2.1}$$

and

$$\begin{aligned}
\mathrm{Cov}(y_i y_j) &= E\left\{\left(y_i - \bar{Y}\right)\left(y_j - \bar{Y}\right)\right\} \\
&= E(y_i y_j) - \bar{Y}^2 \\
&= \frac{1}{N(N - 1)}\left[\left(\sum_{j=1}^{N} Y_j\right)^2 - \sum_{j=1}^{N} Y_j^2 - N(N - 1)\bar{Y}^2\right] \\
&= -S^2/N.
\end{aligned} \tag{2.2}$$

Thus we find that distinct sample members are not independent as in the infinite population case. In fact there is a small negative correlation between the potential sample observations.

The result (2.1) seems to suggest that the divisor N, not $(N - 1)$, would be better in the definition of the finite population variance—for the sake of tidiness. But later results outweigh this, and show that the adopted form, with divisor $(N-1)$, is indeed more convenient.

We can now proceed to study the estimation of the population mean.

2.2 Estimating the mean, $\overline{Y}$

An estimator of $\overline{Y}$, based on a sr sample of size n, with immediate intuitive appeal is the *sample mean*,

$$\bar{y} = \frac{1}{n}\sum_{i=1}^{n} y_i.$$

Let us consider some properties of $\bar{y}$ as an estimator of $\overline{Y}$.

We see that

$$E(\bar{y}) = \overline{Y},$$

for, $E(\bar{y}) = \frac{1}{n} E(y_1 + y_2 + \cdots + y_n) = n\overline{Y}/n = \overline{Y}$. Thus $\bar{y}$ is *unbiased* for $\overline{Y}$.

Also,

$$\text{Var}(\bar{y}) = (1 - f)S^2/n, \tag{2.3}$$

for,

$$\begin{aligned}
\text{Var}(\bar{y}) &= \frac{1}{n^2}\sum_{i=1}^{n} \text{Var}(y_i) + \frac{2}{n^2}\sum_{r<s} \text{Cov}(y_r y_s) \\
&= \frac{1}{n^2}[n(N-1)S^2/N - n(n-1)S^2/N] \\
&= \left(\frac{N-n}{N}\right)S^2/n = (1-f)S^2/n.
\end{aligned}$$

We see in (2.3) the effect of the population being finite. The sampling variance of $\bar{y}$ is reduced by a factor $f = n/N$, the **sampling fraction**, compared with the analogous result for an infinite population. This effect is known as the **finite population correction** (fpc).

If the sampling fraction is small, the fpc has little importance, and we can often ignore it. As a rule of thumb we might ignore the fpc if f is less than about 0.05. The consequence is to slightly over-state the variance of the estimator, $\bar{y}$, but such implied conservatism in any assessment of the accuracy of estimation of $\overline{Y}$ is usually unimportant.

Terminology

The *standard deviation of* $\bar{y}$, $[\text{Var}(\bar{y})]^{1/2}$, will be called its *standard error*.

Thus we find that $\bar{y}$ is *unbiased as an estimator of* $\overline{Y}$, and (2.3) enables us to compare its efficiency with other estimators of $\overline{Y}$ based on simple random samples, or samples obtained from other sampling schemes.

Also, $\bar{y}$ is *consistent* in *the finite population sense*: as $n \to N$, so $\bar{y}$ becomes $\bar{Y}$.

Under appropriate circumstances we can also invoke the approximate *normal* distributional form for $\bar{y}$ to derive confidence intervals for $\bar{Y}$, or to choose a sample size n to meet prescribed accuracy requirements. But before considering these matters let us pose a more basic question.

Within the simple random sampling scheme, how well does $\bar{y}$ compare with other possible estimators of $\bar{Y}$?

One property is easily demonstrated. The sample mean, $\bar{y}$, is the **best linear unbiased estimator** of $\bar{Y}$ based on a simple random sample of size n.

The expression 'best' is used here to mean 'having smallest variance'. The result does not imply any global optimality among all estimators, $\tilde{\theta}(y)$, of $\bar{Y}$, but it is useful to know that in the class of easily calculated *linear unbiased estimators*, the simple form $\bar{y}$ is best.

Let us confirm that this is so. Consider any linear unbiased estimator,

$$t = \sum_{i=1}^{n} a_i y_i,$$

with the condition

$$\sum_{i=1}^{n} a_i = 1$$

imposed to ensure unbiasedness. So

$$\text{Var}(t) = \left(\frac{N-1}{N}\right) S^2 \sum_{i=1}^{n} a_i^2 - \frac{2S^2}{N} \sum_{r<s} a_r a_s = S^2 \left(\sum_{i=1}^{n} a_i^2 - \frac{1}{N} \right).$$

Thus we need

$$\sum_{i=1}^{n-1} a_i^2 + \left(1 - \sum_{i=1}^{n-1} a_i \right)^2$$

to be a minimum, which will be so if

$$a_i = 1 - \sum_{i=1}^{n-1} a_i = a_n.$$

In other words a_i must be constant ($i = 1, 2, ..., n$), and the unbiasedness condition shows that, for minimum variance,

$$a_i = 1/n,$$

yielding the sample mean, $\bar{y}$, as the *best linear unbiased estimator*.

2.3 Random sampling with replacement

Sampling with replacement from a finite population does not feature widely in sample survey work. Nonetheless, it is interesting to note how the results would differ if we chose a sample of size n by picking each sample member *at random* but *with replacement*, from the overall population of size N.

Now, each observation has probability N^{-1} of being any of the population members (rather than successive observations having probabilities N^{-1}, $(N-1)^{-1}$, $(N-2)^{-1}$ ... of being any of the *non-observed* population members).

Effectively, we have reverted to the conventional statistical method where we are drawing a random sample of size n from a discrete uniform distribution on the set of values $(Y_1, Y_2, ..., Y_N)$.

So the sample mean, $\bar{y}$, is unbiased for the mean of the distribution, which is again $\bar{Y} = (\sum_{i=1}^{N} Y_i)/N$. Furthermore, it has sampling variance σ^2/n, where σ^2 is the variance of the uniform distribution which takes the form

$$\sigma^2 = \frac{1}{N} \sum_{i=1}^{N} (Y_i - \bar{Y})^2 = (N-1)S^2/N.$$

So

$$\text{Var}(\bar{y}) = \left(1 - \frac{1}{N}\right) S^2/n$$

compared with $[1 - (n/N)]S^2/n$ in the case of simple random sampling (*without replacement*).

Thus *sampling with replacement is bound to be less efficient*: the relative efficiency is $(N-n)/(N-1)$. See also Section 2.9 below.

2.4 Estimating the variance, S^2

The expression (2.3) for $\text{Var}(\bar{y})$ will need to be used in three possible ways:

    (i) to assess the precision of the estimator $\bar{y}$,
   (ii) to compare $\bar{y}$ with other estimators of $\bar{Y}$,
  (iii) to determine the size of sample needed to yield a desired precision.

Typically, however, we will not know the value of S^2, so that to make use of the sampling variance (2.3) we must estimate S^2 from sample data. Using the simple random sample $y_1, y_2, ..., y_n$ we might try (cf. infinite populations) using

$$s^2 = \frac{1}{n-1} \sum_{i=1}^{n} (y_i - \bar{y})^2.$$

It turns out that

$$E(s^2) = S^2,$$

for,

$$E(s^2) = \frac{n}{n-1} \left[\sum_{j=1}^{N} Y_j^2/N - E(\bar{y}^2) \right]$$

$$= \frac{n}{n-1}\left[\sum_{j=1}^{N} Y_j^2/N - (1-f)S^2/n - \overline{Y}^2\right]$$

$$= \frac{n}{n-1}\left(\frac{N-1}{N} - \frac{1-f}{n}\right)S^2$$

$$= S^2.$$

So that s^2 is unbiased for S^2.

In relation to problems (i) and (ii), above, we can substitute for the unknown population variance in (2.3) the unbiased sample estimator, s^2, and we have an *unbiased estimator of* Var($\overline{y}$) as

$$s^2(\overline{y}) = (1-f)s^2/n.$$

Occasions may also arise where estimation of S^2 is of interest in its own right; again s^2 serves for this purpose.

But the problem (iii), where we wish to determine the size of sample needed to achieve a desired precision, is less straightforward if S^2 is not known. This is because the sample estimator s^2 is now of no relevance since we do not have a sample from which to extract it! We need to determine the required sample size *prior to sampling*, and we shall consider later in Section 2.6 how to face up to the difficulty of an unknown population variance.

2.5 Confidence intervals for $\overline{Y}$

Example 2.2

Suppose that for the *Global Games* team data described in Section 1.7 we decide to estimate the mean weight $\overline{Y}$ from the sample mean of a simple random sample of size 5 (see Example 2.1 above). Thus if our sample is 76, 54, 76, 71, 57 the estimate of $\overline{Y}$ is

$$\overline{y} = 66.80$$

From (2.3) we see that the sampling variance of the sample mean is

$$(1 - 5/25)S^2/5$$

where the population variance for the Y-values is in fact 73.95. Thus the standard error of $\overline{y}$ is $0.4S = 3.44$. In fact, $\overline{Y} = 67.64$ so that our estimate $\overline{y} = 66.80$ is very close: only about 0.24 standard errors less than the true value. But in general, of course, it need not have been as close as this!

But what of the actual distribution of $\overline{y}$? We can obtain some indication of its form from taking repeated sr samples of size five. One thousand such samples generated on a computer yielded values of $\overline{y}$ represented

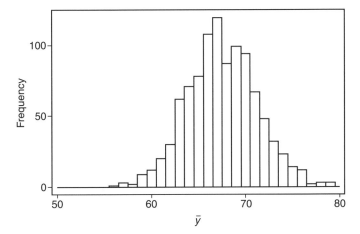

Fig. 2.1. Histogram of 1000 values of $\bar{y}$ from sr samples of size 5 in the *Global Games* team data

by the histogram* shown in Figure 2.1. The sample variance of the 1000 sample means is 13.61, which compares well with the expected value 13.31. (The average of the 1000 sample means is 67.53.)

We see in Figure 2.1 that, even for such a small sample size and non-negligible sampling fraction, the sampling distribution of $\bar{y}$ shows no substantial lack of symmetry. But this effect is assisted by the fact that the finite population of Y-values itself shows relatively little 'skewness'. This will not always be true of course.

We will frequently encounter skew populations in practice, often positively skew in the sense of exhibiting a long tail of large Y-values. Consider, for example, the numbers of children in different families, or family annual incomes. Even more extreme situations will be encountered, with a maximum frequency at the lowest Y-values so that we have an almost *i*-shaped distribution of population values. Consider, for example, the numbers of claims on an insurance policy arising from a population of insured people, or the number of times a medical practitioner sees each of the patients on his panel over a given year.

But, supported by a finite population analogue of the *Central Limit Theorem*, it can often be assumed that the sr sample mean has approximately a *normal sampling distribution*. That is,

$$\bar{y} \sim N(\bar{Y}, (1-f)S^2/n). \tag{2.4}$$

This assumption is often reasonable even in the presence of skewness in the population. A rough guide has been proposed for positive skew populations requiring that the sample size n should satisfy

$$n > 25G_1^2$$

---

* For this histogram and for the others throughout the book, the data have been grouped in intervals of width 1 kg. The ordinate label 'frequency' relates throughout to the numbers of observations in such equal-sized intervals.

where

$$G_1 = \sum_{i=1}^{N} (Y_i - \bar{Y})^3 / NS^3$$

(G is the finite-population analogue of Fisher's *coefficient of skewness*). In addition, the sampling fraction, f, should not be too large. Much discussion of the propriety of the normal approximation (2.4) has appeared in the literature. Cochran (1977, Chapter 2) gives further details and relevant references. See also Thompson (1992, Chapter 3) which includes discussion of the Central Limit Theorem for finite populations.

Where appropriate we can use the normal distribution to make further inferences about $\bar{Y}$. We might wish to construct a confidence interval for $\bar{Y}$. An approximate $100(1 - \alpha)\%$ *symmetric two-sided confidence interval* for $\bar{Y}$ can be written

$$\bar{y} - z_\alpha S \sqrt{(1 - f)/n} < \bar{Y} < \bar{y} + z_\alpha S \sqrt{(1 - f)/n}, \tag{2.5}$$

where z_α is the double-tailed α-point of $N(0, 1)$. That is, if $Z \sim N(0, 1)$, then $\Pr(|Z| > z_\alpha) = \alpha$.

But S^2 will not be known in practice. Replacing S^2 by the sample estimate, s^2, will be reasonable, provided n is sufficiently large. By analogy with the infinite population case, it is sometimes claimed (e.g. Cochran, 1977, Section 2.8) that allowance can be made for not knowing S^2 by using *Student's t-distribution* rather than the normal distribution, when n is small (less than about 40, say).

The use of the t-distribution does not, for finite populations, have the formal justification it does with infinite population sampling of a normal random variable. It is used in finite population sampling as an empirical approximation but the population needs to be fairly symmetric.

Proceeding with the use of the t-distribution then, we have an approximate $100(1 - \alpha)\%$ *symmetric two-sided confidence interval for $\bar{Y}$* in the form

$$\bar{y} - t_{n-1}(\alpha)s\sqrt{(1 - f)/n} < \bar{Y} < \bar{y} + t_{n-1}(\alpha)s\sqrt{(1 - f)/n} \tag{2.6}$$

where $t_{n-1}(\alpha)$ is the double-tailed α-point of t with $(n - 1)$ degrees of freedom. Values of z_α, and $t_{n-1}(\alpha)$ for a range of values of n are given in Table 2 of the Appendix, for $\alpha = 0.1, 0.05, 0.02, 0.01, 0.002$ and 0.001. If n is more than about 40, the normal distribution percentage points will usually be reasonable.

Since sample surveys commonly relate to very large populations (say $N = 10\,000$ or more) with substantial sample sizes (say $n = 100$ or more), we will frequently be safe in adopting the form (2.5), replacing S by s, without regard to the fine details of justifying the *normal* distributional form for $\bar{y}$ or serious concern for the sampling fluctuations of s as an estimate of S.

However, one word of warning is needed on this latter issue. The sampling variance of s^2 is highly sensitive to the value of the *fourth* central moment of the Y-values in the finite population. Typically, the larger this fourth moment the larger is $\text{Var}(s^2)$. This is particularly serious when we attempt to compare the precision of alternative estimators and need to estimate S^2. It can be even more serious when we need to choose a sample size to yield a required precision (as described in the next section).

Example 2.3

In a particular sector of industry a survey was conducted in an attempt to investigate the extent of absenteeism not connected with illness or official holidays. A random sample of 1000 workers out of a total workforce of 36 000 were asked how many days they have taken off work, in the previous six months, as 'casual holidays'. The results were as follows.

Days off	0	1	2	3	4	5	6	7	8	9
No. of workers	451	162	187	112	49	21	5	11	2	0

To estimate the average number, $\bar{Y}$, of days 'casual holiday' taken by workers in the industry we can use the sample mean

$$\bar{y} = 1.296.$$

The sample variance is

$$s^2 = 2.397.$$

Using the normal approximation to the distribution of the sample mean we can obtain an approximate 95% symmetric two-sided confidence interval for $\bar{Y}$ as

$$1.201 < \bar{Y} < 1.391$$

[or, $1.200 < \bar{Y} < 1.392$ (ignoring the fpc)].

Note some characteristic features of this problem. The distribution of values in the population is obviously highly skew. Such skewness would in general affect the propriety of the normal approximation, although the current sample size of 1000 is an adequate safeguard.

Note that, beyond the statistical considerations, there is the inevitable problem of assessing the accuracy of the information given on such a controversial issue. (Promises of confidentiality may not necessarily allay all concern!). If the survey involved compulsory response, fears for its accuracy become more acute. Suppose that a number of those questioned just refused to answer. How would we deal with such 'non-response': might it yield an unrepresentative sample? Perhaps the most serious offenders are most likely not to respond. How would we then process the data to estimate $\bar{Y}$? We will consider the non-response problem further in Section 6.6.2.

2.6 Choice of sample size, *n*

Clearly an increase in sample size will lead to an increase in the precision of $\bar{y}$ as an estimator of $\bar{Y}$—but the sampling costs will also typically increase and there is likely to be some limit on what we can afford. Too large a sample will imply a waste of resources; too small a sample is likely to produce an estimator of inadequate precision.

Ideally we should state the precision we require, or the maximum cost which can be expended, and choose the sample size accordingly.

Such an aim involves a complex array of considerations. What is the cost structure for sampling in a given situation; how do we assess the precision we require of our estimators; how do we balance needs in relation to *different* population characteristics which may be of interest; how do we deal with lack of knowledge of unknown parameters (e.g. the population variance) which may affect the precision of estimators?

We will consider only one simple situation. We assume that the object is to estimate a single characteristic, the population mean $\bar{Y}$, by using a sr sample mean $\bar{y}$, *restricting to an acceptable level the probability that the absolute difference between $\bar{Y}$ and $\bar{y}$ is greater than some specified value.*

No direct consideration of costs arises here. However, if it happens that sampling costs are directly proportional to sample size, it turns out that we achieve our aim for minimum cost.

Thus suppose we seek the minimum value of n that ensures that

$$\Pr\left(|\bar{Y} - \bar{y}| > d\right) \leq \alpha \tag{2.7}$$

for some prescribed d and (small) α. The sampler needs to specify the tolerance d, and the risk α of not obtaining an estimate within such tolerance.

We can rewrite (2.7) as

$$\Pr\left(\frac{|\bar{Y} - \bar{y}|}{S\sqrt{(1-f)/n}} > \frac{d}{S\sqrt{(1-f)/n}}\right) \leq \alpha, \tag{2.8}$$

so that, using the normal approximation (2.4), we require

$$\frac{d}{S\sqrt{(1-f)/n}} \geq z_\alpha$$

or,

$$n \geq N\left[1 + N\left(\frac{d}{z_\alpha S}\right)^2\right]^{-1}. \tag{2.9}$$

Equivalently, (2.8) declares that

$$\mathrm{Var}(\bar{y}) \leq (d/z_\alpha)^2 = V \quad \text{(say)}.$$

The inequality (2.9) can be written

$$n \geq S^2/V\left[1 + \frac{1}{N}S^2/V\right]^{-1},$$

so that as a *first approximation* to the required sample size, we could take

$$n_0 = S^2/V \tag{2.10}$$

provided $(S^2/V)/N$ is small.

This is an *overassessment*, but it will be reasonable unless the provisional sampling fraction, n_0/N, is substantial. If this is so, we would need to reduce our assessment of the *required sample size* to

$$n_0(1 + n_0/N)^{-1}.$$

This presupposes, however, that S^2 is known. Such a possibility is remote in practice and thus we must face the more difficult task of estimating the minimum sample size required to satisfy (2.7) when S^2 is unknown. There are basically four ways in which we might try to do this.

(i) From pilot studies

Often a *pilot study* may be conducted prior to a major sample survey (see Section 6.2). This can serve a variety of purposes including the comparison of different sampling frames and the examination of any implicit practical difficulties which may be encountered in the sampling procedure.

If such a pilot study itself takes the form of a simple random sample, its results may give some indication of the value of S^2 for use in the choice of the sample size of the main survey; if the pilot sample is not obtained by a probability sampling procedure we must be circumspect in such an application of the results. For convenience, a pilot study is often restricted to some limited part of the population. If so, the estimate of S^2 which it yields can be quite biased.

(ii) From previous surveys

It is not uncommon to find that other surveys have been conducted elsewhere or at different times which have studied *similar* characteristics in *similar* populations. This is particularly common in educational, medical, or sociological investigations—it may just be that a different age-group of pupils, or results for a different year, or effects in a different city or social structure, are being considered. Often the measure of variability from earlier surveys can be used to estimate S^2 for the present population, in order to choose the required sample size to meet any prescription of precision in the current work. But again precautions must be taken in extrapolating from one situation to another.

(iii) From a preliminary sample

This is the most reliable approach, but it may not be feasible on administrative or cost considerations. It operates as follows. A preliminary sr sample of size n_1 is taken and used to estimate S^2 by means of the sample variance s_1^2. We aim to ensure that n_1 is inadequate to achieve the required precision, and then to augment the sample with a further sr sample of size $(n - n_1)$, where $(n - n_1)$ is chosen by using s_1^2 as the necessary preliminary estimate of S^2.

A detailed study of this procedure shows that, under reasonable conditions, the total sample size, *ignoring the fpc*, needs to be

$$(1 + 2/n_1)s_1^2/V$$

an essential increase by the factor $(1 + 2/n_1)$ over what would be needed if S^2 were known.

This approach, if feasible, is undoubtedly the most objective and reliable. Further details are given by Cochran (1977, Section 4.7). See also Hedayat and Sinha (1991, Section 4.6); original work on this theme is due to Cox (1952).

(iv)   From practical considerations of the structure of the population

Occasionally we will have some knowledge of the structure of the population which throws light on the value of S^2. Suppose we are considering

    (a) the numbers of misprints in books (of roughly the same size, or over a pre-scribed number of pages) issued by a particular publisher over a certain period of time, or

    (b) the number of faults that occur in video-recorders of a particular type in the first year of their use.

In both cases there is reason to believe that the Y-values might vary roughly in the manner of a Poisson distribution, so that it is plausible to assume that S^2 is of the same order of magnitude as $\bar{Y}$. Any information we have about the possible value of $\bar{Y}$ (for example, from other similar studies) can then be used to approximate S^2 and assist in the choice of the required sample size.

This is equivalent, of course, to extending our method of inference from *randomisation inference* to *model-based* (or at least model-assisted) *inference* in the sense of the distinction drawn in Section 1.6 and elaborated in Section 2.14 below.

Returning to the Poisson model, where we can assume that $S^2 = \bar{Y}$, then we can obtain an approximate $100(1 - \alpha)\%$ symmetric two-sided confidence interval for $\bar{Y}$ directly, without the need for an estimate of variability. Using the normal approximation to the Poisson distribution we have

$$\Pr[|\bar{Y} - \bar{y}| < z_\alpha \sqrt{\bar{Y}(1 - f)/n}] = 1 - \alpha.$$

Thus, the confidence interval is obtained in the form

$$\bar{Y}^2 - [2\bar{y} + z_\alpha^2(1 - f)/n]\bar{Y} + \bar{y}^2 < 0$$

i.e. *as the region between the two roots of the equation*

$$\bar{Y}^2 - [2\bar{y} + z_\alpha^2(1 - f)/n]\bar{Y} + \bar{y}^2 = 0.$$

Then again, if we are interested in estimating a proportion P we shall see that the sampling variance of the sr sample estimator is simply related to P. Reasonable bounds can be placed on this variance to obtain some idea of the required sample size. We shall return to this point later (Section 2.12).

2.7   **Systematic sampling**

Suppose we wish to draw a sample of size n from a population of size N and have available a *complete list* of the population members. The list is of great value in specifying the sample we intend to draw. Members of the sample can be identified on the list, and then sought in the population. Thus we might sample the students in a university by using a published list of students for that university; or the books in a library by going through the catalogue of books held by the library (multiple entries in such indexes raise interesting sampling problems!). The list might be in printed form or stored electronically on a computer or Website.

Even with such a complete list, however, the choice of a random sample can be tedious and time-consuming. Imagine choosing 500 students at random from a list of 8500. There is a great temptation to seek an easy way out, and one method of doing so, which is commonly employed, is to take a **systematic sample**.

The principle is that sample members are chosen in a regular manner working progressively through the list. Consider the student example; since $8500 = 500 \times 17$, we might choose a student *at random* in the first 17 on the list, and then take every 17th student subsequently. This is a *systematic sample* with sampling interval 17. Note that it is not a strictly random sample in view of the deterministic method of choice of sample members after the first one.

Whether or not N is a multiple of n, the method is readily described. Suppose we take a sampling interval, k, and we have $N = (n - 1)k + t$ where $0 < t \leqslant k$. We pick an individual at random from the first t and then choose each subsequent kth member on the list, to obtain a systematic sample of size n, where n is the least integer greater than or equal to N/k.

Such a sampling method is clearly very easy to carry out, which is a great advantage, particularly in frequently repeated surveys, or when the sample is chosen 'on site' and a list is available. Sometimes even the limited randomisation, in the choice of the first member, is dispensed with, and the sequence is fully prescribed at the outset. For example, it might be decided to take the 9th student, and every 17th one subsequently, on the basis that this seems to be a systematic policy.

What are the likely effects of such a commonly used sampling approach?

Administrative convenience is a clear benefit. We do not even need to know the population size N; we can merely specify a sampling fraction $1/k$.

Apart from the saving in time and effort, there is also some sort of intuitive appeal in systematic sampling: it seems to 'span the population' in a way that might lead to more 'representative' results than those obtained from random choice.

There are obvious dangers however. The procedure is not sr sampling since not all possible samples of size n have an equal chance of occurring. Related to this is the fact that we cannot readily obtain a variance estimate from such a single sample.

If, however, there is no obvious order pattern (it is quasi-random) we can act as if the sample is effectively a sr sample and thus couple the benefits of such a sampling principle with the convenience of systematic choice.

But order can be important. Consider a medical trial in which patients have been listed in order of treatment: three forms of treatment, A, B and C, having been given in rotation to successive patients. If we now take a systematic sample, with sampling interval k which is a multiple of 3, we will only obtain patients *with a specific single treatment regime* (A or B or C). Thus we obtain a very unrepresentative sample. Such an extreme problem may not frequently arise, but less deterministic order patterns are not uncommon, e.g. trends (the listed members get successively older), groups (different parts of the list relate to different geographic regions) or even cyclic variations (sales data for a product in time order).

All prospects in relation to sr sampling are possible. *We can do better*: for example, geographical groupings may *improve* representatively–we are then effectively taking a *stratified sample* (see Chapter 4). *We could do much worse*, as in the medical trial example above. Or we can effectively be taking just a sr sample and this is the assumption commonly made in the absence of obvious counter-indications.

We shall see (in Section 5.2 below) that systematic sampling can be represented as a specific case of *one-stage cluster sampling* and as such we can determine some of the formal properties of estimators obtained from such a systematic approach. In particular we will find that the systematic sample mean $\bar{y}_s$ is essentially unbiased for $\bar{Y}$ with variance

$$\mathrm{Var}\left(\bar{y}_s\right) = \frac{n}{N} \sum_{i=1}^{N/n} \left(\bar{Y}_i - \bar{Y}^2\right)$$

where $\bar{Y}_i$ is the mean of the *i*th potential systematic sample which might have been drawn, from the N/n possible samples. But see section 5.2 for more details.

Example 2.4

Suppose we estimate the mean weight of the *Global Games* team using systematic sampling. A systematic sample of size five (with one member picked at random from the first 10 on the list shown as Table 1.2 and every tenth subsequently chosen to make up the sample) yields an estimate

$$\bar{y}_s = 67.68$$

compared with the population value, $\bar{Y} = 67.64$.

Only 10 possible systematic samples could be obtained in this way. They are all equally likely to arise and we can readily calculate $\mathrm{Var}(\bar{y}_s)$. It takes the value 2.128.

The distribution of the different possible values (grouped as for other estimators) is shown in Figure 2.2.

Thus, $\bar{y}_s$ *appears to be* highly efficient—much more so than $\bar{y}$ (for $n = 5$ we found $\mathrm{Var}(\bar{y}) = 13.61$ in Example 2.2). Indeed, we will not find such a small sampling variance even for any of the more structured methods below (such as ratio and regression estimators or estimators based on stratification or clustering).

So are we to conclude that systematic sampling is 'super-efficient'? For the population as ordered in Table 1.2, any systematic sample *must contain* three male and two female members (proportional to their incidence

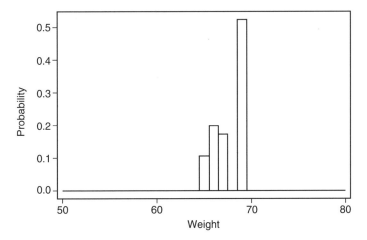

Fig. 2.2. Sampling distribution of systematic sample means for samples of size 5 in the *Global Games* data.

in the overall population) and the possible $\bar{y}_s$ values are clustered together (from about 65.5 to 69.5). But suppose the population was differently ordered, either (a) keeping the male and female members in their respective blocks or (b) reordering over the whole population. We then find very different apparent sampling behaviour for $\bar{y}_s$.

The values of $\text{Var}(\bar{y}_s)$ for $n = 5$ and for 10 *different* random reorderings under (a) and (b) range very widely, as follows:

(a) 4.91	4.85	11.74	8.00	4.65	8.70	4.41	5.76	4.28	14.53
(b) 11.91	13.69	1.85	6.21	8.64	10.97	16.78	8.05	18.66	23.34

So we see that the initially observed $\text{Var}(\bar{Y}_s) = 2.218$ seems to have arisen from a particularly fortunate ordering of the population (and the value of 13.61 previously observed for $\text{Var}(\bar{Y})$ is in the midst of the values obtained for the different orderings).

Thus it seems that we cannot infer how well systematic sampling will perform (at least in such a small population); it is possible that it *might* produce a low-variance estimator (e.g. with $\text{Var}(\bar{y}_s) = 2.218$ in this example) or it *might not* (see $\text{Var}(\bar{y}_s) = 23.34$ in the display above) depending on the order in which the population members are listed.

2.8   Estimating the population total, Y_T

There are many situations in which we are interested in estimating the population total

$$Y_T = N\bar{Y}$$

rather than the population mean $\bar{Y}$. For example, in a survey of annual yields of wheat for farms in Ohio, the concern may be to estimate the state's total annual wheat yield. In view of the simple relationship between Y_T and $\bar{Y}$, no substantial extra difficulties arise; we can immediately extend the results we have obtained concerning the estimation of $\bar{Y}$.

The sr sample estimator of Y_T which is commonly used is

$$y_T = N\bar{y},$$

and the earlier results confirm that y_T is unbiased for Y_T; that is,

$$\boxed{E(y_T) = Y_T}$$

and that

$$\boxed{\mathrm{Var}(y_T) = (1 - f)N^2 S^2/n} \qquad (2.11)$$

Furthermore, y_T is *the minimum variance linear unbiased estimator of Y_T based on a simple random sample of size n.*

With similar qualifications as before concerning the sample size, n, and value of the sampling fraction, f, we can use the normal approximation

$$y_T \sim N[Y_T, \ (1 - f)N^2 S^2/n]$$

to construct confidence intervals for Y_T, or to choose a sample size to meet specified requirements concerning the precision of estimation of Y_T.

For example, if n is more than about 40, an approximate $100(1 - \alpha)\%$ *symmetric two-sided confidence interval* for Y_T is given by

$$y_T - z_\alpha NS\sqrt{(1 - f)/n} < Y_T < y_T + z_\alpha NS\sqrt{(1 - f)/n}.$$

For smaller n, the use of percentage points $t_{n-1}(\alpha)$ for the t-distribution with $(n - 1)$ degrees of freedom, in place of z_α, is claimed to be preferable.

Let us consider the question of choosing n to ensure that

$$\Pr(|\, y_T - Y_T\,| > d) \leq \alpha.$$

Using the normal approximation, this requires

$$n \geq N\left[1 + \frac{1}{N}\left(\frac{d}{z_\alpha S}\right)^2\right]^{-1}. \qquad (2.12)$$

Equivalently, we require

$$\mathrm{Var}(y_T) \leq (d/z_\alpha)^2 = V,$$

so that (2.12) becomes

$$n \geq \frac{N^2 S^2}{V}\left(1 + \frac{1}{N}\frac{N^2 S^2}{V}\right)^{-1}.$$

So if NS^2/V is very much less than 1, it will be reasonable to take

$$n_0 = \frac{N^2 S^2}{V}$$

as the required sample size; otherwise we must use

$$n_0(1 + n_0/N)^{-1}.$$

In some respects it is more natural to express the accuracy we require of an estimator of Y_T (or even of $\bar{Y}$) in proportional, rather than absolute, terms. Thus we might ask what sample size is needed to ensure that

$$\Pr(|y_T - Y_T| > \xi Y_T) \leqslant \alpha. \tag{2.13}$$

For example we may want to be at least 95% certain that y_T is within 2% of Y_T. Then $\alpha = 0.05$, $\xi = 0.02$.

But this is less straightforward. It becomes necessary to replace d in (2.12) by ξY_T, and the appropriate value of n now depends on Y_T, the unknown quantity we are trying to estimate. The best we can hope for is to obtain a rough estimate of the sample size necessary to satisfy (2.13), by replacing Y_T in the right hand side of the inequality by some provisional estimate (possibly based on earlier surveys or experience).

For example, if the y-values have an approximate Poisson distribution (so that $S^2 = \bar{Y} = Y_T/N$) then (2.12) takes the form

$$n > N\left[1 + \left(\frac{\xi}{z_\alpha}\right)^2 Y_T\right]^{-1}.$$

Example 2.5

To obtain an early indication of the total sales of Christmas cards throughout a network of 243 retail stationery shops, it is decided that a random sample of the shops should submit returns of their card sales by the end of January. How large a sample is needed to estimate total sales to within 10% of the correct figure with 95% assurance?

By March of each year, precise figures of total sales of cards at all shops are available. For the previous three years the number of shops in the network has remained much the same; the total card sales, and standard deviations of sales from shop to shop, have been (in units of 10 000 cards)

Y_T	S
321.7	0.826
366.8	0.776
401.0	0.804

So for the current year we might reasonably expect that Y_T and S will be of the order of 420 (reflecting annual growth) and 0.8, respectively. To obtain the required precision from the January returns it will consequently be necessary to take a simple random sample of size n, where, from (2.12),

$$n \geqslant 243\left[1 + \frac{1}{243}\left(\frac{0.10 \times 420}{0.8 \times 1.96}\right)^2\right]^{-1}$$

$$= 243(1 + 2.96)^{-1}$$

$$= 61.48.$$

So a sample of size $n = 62$ is needed.

Here $n_0 = N^2 S^2 / V = 82.30$, so that $n_0/N = 0.34$. So the first approximation would not have been good enough. We do in fact need to use the more accurate expression $n_0(1 + n_0/N)^{-1}$ to obtain the above value of 62 for the required sample size.

Suppose that such a sample of size 62 yields an estimate

$$y_T = 427.4,$$

then we obtain an approximate 95% confidence interval for Y_T as

$$385.6 < Y_T < 469.2,$$

reflecting (roughly) the 10% absolute accuracy we sought.

When attempting to estimate Y_T with prescribed precision, we again typically encounter difficulties because we are unlikely to know the value of S^2, so that some preliminary indication of its value will be needed using one of the methods (i), (ii), (iii), or (iv) described in Section 2.6. Example 2.5 illustrates method (ii), in a case where the previous information arises from a complete enumeration, rather than from a sample survey, of a similar situation.

Finally we should note that the population size N *needs to be known* if we are to use y_T to estimate Y_T or to assess the sampling behaviour of y_T. This is true also for the study (in Section 2.9) of the Hansen–Hurwitz estimator $\widetilde{\overline{y}}$ as an estimator of $\overline{Y}$. In most situations N will be known, or can be estimated with fair accuracy. Where this is not so, difficulties arise even in the very choice of a sr sample.

2.9 Non-epsem sampling

In *epsem* sampling, of which sr sampling is an example, we use a principle of *selecting sampling units with equal probability*. Sometimes, however, this is not straightforward nor necessarily the best approach.

Consider the example of estimating the total wheat yield Y_T for wheat farms in Ohio. If we could list all the farms, we could take a sr sample and use the methods described above. But suppose that we marked points at random on a map of Ohio to identify our sample of n farms, and further assume that this had the effect of selecting the farms with *probabilities proportional to their land areas* X_j, since obviously the larger farm would be more likely to be pin-pointed. This approach is *not* epsem.

There could also be a strong correlation between the wheat yields Y_j and the areas X_j. The extreme prospect is that $Y_j = kX_j$: that is, wheat yield is directly proportional to area and we would now be conducting what is termed **pps sampling**: sampling with probability proportional to size (of sampled variable). If the Y_i and X_i are positively correlated but not strictly proportional we would be sampling with probability proportional to estimated size (**ppes sampling**).

Let us examine the effect of such an approach (and prospect), if we obtain a sample $y_1, y_2, \ldots, y_n$ chosen *without replacement* (to simplify the algebra) and wish to estimate the total wheat yield Y_T. Consider the estimator

$$\tilde{y}_T = \frac{X_T}{n} \sum_{j=1}^{N} \left(\frac{y_j}{x_j} \right)$$

assuming that the total area, X_T, is known and that we also observe the sample $x_1, x_2, \ldots, x_n$ for the sampled farms.

For a typical wheat yield y and area x chosen by sampling with probability proportional to x (ppes sampling), we have

$$E\left(\frac{y}{x} \right) = \frac{1}{X_T} \sum_{j=1}^{N} \left(\frac{Y_j}{X_j} \right) X_j = Y_T / X_T.$$

Thus

$$E(\tilde{Y}_T) = \frac{X_T}{n} n \frac{Y_T}{X_T} = Y_T,$$

so that $\tilde{Y}_T$ is unbiased for Y_T

Furthermore,

$$\mathrm{Var}\left(\frac{y}{x} \right) = E\left[\left(\frac{y}{x} \right)^2 \right] - Y_T^2 / X_T^2 = \frac{1}{X_T} \sum_{j=1}^{N} \left(\frac{Y_j^2}{X_j} \right) - Y_T^2 / X_T^2$$

Thus

$$\mathrm{Var}\left(\tilde{Y}_T \right) = \frac{X_T^2}{n^2} \left\{ n \left[\frac{1}{X_T} \sum_{j=1}^{N} \left(\frac{Y_j^2}{X_j} \right) - Y_T^2 / X_T^2 \right] \right\} = \frac{X_T}{n} \sum_{j=1}^{N} \left(\frac{Y_j}{X_j} \right)^2 - Y_T^2 / n.$$

In the extreme case, $Y_j = kX_j$ and $Y_T = kX_T$. Hence

$$\mathrm{Var}(\tilde{Y}_T) = k^2 X_T^2 / n - k^2 X_T^2 / n$$
$$= 0$$

and we have discovered *the perfect estimator, $\tilde{Y}_T$, of Y_T with zero variance!*

Of course this is an artificial example of no practical value! If $Y_T = kX_T$, and we know X_T, then we also know Y_T without the need for estimation. But an important principle is embedded in this example. If the Y_j and X_j are correlated, albeit imperfectly and short of full proportionality, *ppes sampling may still be efficient and improve on sr sampling*.

Also knowledge of X_T may still be critically important. We shall return to these themes, in Chapter 3 on *ratios* and allied topics, where we examine for sr sampling the value of simultaneously observing the principal measure Y and a correlated measure X and in Chapters 3 and 5 where we further consider *pps* or *ppes sampling*, including the prospect of *sampling with replacement*.

When we move away from sr sampling, there are a number of complications that arise in seeking to employ an arbitrary sampling scheme $\{S, \pi\}$. The first is how to generate the samples $s_i \in S$ with their corresponding probabilities $\pi_i \in \pi$ The second is how to derive the statistical properties of estimators based on the scheme $\{S, \pi\}$.

To introduce this topic, consider first a sampling scheme which can be defined in terms of sequential generation of successive sample members, especially one in which population members are chosen *with replacement*. A general scheme of this type arises when, at each stage, the different values $Y_1, Y_2, ..., Y_N$ can occur independently with respective probabilities $p_1, p_2, ..., p_N$. Suppose we so choose a sample of size n, the sample values being $y_1, y_2, ..., y_n$.

Again we might estimate $\bar{Y}$ by the simple average

$$\tilde{y} = \frac{1}{n}\sum_{i=1}^{n} y_i \text{ or, correspondingly, } Y_T \text{ by } N\tilde{y}.$$

The properties of $\tilde{y}$ are easily determined since $y_1, y_2, ..., y_n$ is a random sample from a known probability distribution. We have

$$E(\tilde{y}) = \mu = \sum_{i=1}^{N} p_i Y_i$$

and

$$\text{Var}(\tilde{y}) = \frac{1}{n}\left\{\sum_{i=1}^{N} p_i Y_i^2 - \mu^2\right\}.$$

But μ will *not* equal $\bar{Y}$ except in special circumstances (e.g. when $p_i = 1/n$, $i = 1$, $2, ..., n$), so that $\tilde{y}$ is in general biased and can have large expected mean square error.

The situation is improved if we use a different estimator in which the observations are divided by N times their respective probabilities of occurrence, so that we use

$$\tilde{\tilde{y}} = \frac{1}{n}\sum_{i=1}^{n}(y_i/Np_i), \tag{2.14}$$

where p_i is that member of the set $\{p_1, p_2, ..., p_N\}$ corresponding to the population value chosen as the ith member of the sample. *Equivalently*, we are now using the sample mean of a random sample drawn with replacement from a distribution in which values $Y_i = Y_i/Np_i$ arise with probabilities $p_i(i = 1, 2, ..., N)$. The mean and variance of this distribution are, of course, $\bar{Y}$ and

$$\frac{1}{N^2}\sum_{i=1}^{N} Y_i^2/p_i - \bar{Y}^2,$$

respectively. Thus $\widetilde{\widetilde{y}}$ is now *unbiased* for $\bar{Y}$, with variance

$$\frac{1}{nN^2}\left\{\sum_{i=1}^{N}\left(Y_i^2/p_i\right)-(N\bar{Y})^2\right\}.$$

This variance can again in principle be arbitrarily small; if we could choose $p_i = Y_i/N\bar{Y}$, then each sampled value is precisely $\bar{Y}$ and $\text{Var}(\widetilde{\widetilde{y}}) = 0$! This is again pps sampling. The estimator (2.14) is known as the **Hansen–Hurwitz estimator** (Hansen and Hurwitz, 1943)—note that specific population members may be included more than once in the sample; cf the Horvitz–Thompson estimator below.

Whilst this is *again* of no immediate practical use, of course, since it implies a knowledge of the precise value of $\bar{Y}$ in which case we would not need to sample the population, this pathological result does motivate a particular style of sampling which can often be used with advantage. It can be particularly useful to adopt a non-epsem scheme on occasions, with prescribed selection probabilities p_i and with-replacement sampling.

The optimum aim is to sample the population with the probabilities of selection of different members *proportional to their values*, Y_i. Without any auxiliary information about the population being sampled, this is unrealistic. But with some knowledge of the population structure, or the facility for sampling an auxiliary variable, X, correlated with Y, progress in this direction is possible. As examples, we are led to consider in this respect *sampling with replacement* for ratio estimation, or in cluster sampling. Some further details are given in Chapters 3 and 5 below.

The estimator (2.14) is a special case of an important general class of estimators. Suppose we are sampling according to a scheme $\{\mathbf{S}, \boldsymbol{\pi}\}$ in which we obtain samples s_j where individual population values Y_i enter the sample with probability p_i ($i = 1$, $2, \dots, N$) and pairs (Y_i, Y_j) arise with probability p_{ij}.

Note the generality of this scheme.

If $p_{ii} \neq 0$ we allow for the possibility of sampling with replacement. The $s_i \in S$ do not have to be *n*-tuples (samples of *n* members of $Y_1, Y_2, \dots, Y_N$) but there is no reason why they should not be.

Suppose we wish to estimate the population total Y_T using the estimator

$$y_{HT} = \sum_{i \in s_j} y_i \big/ p_i = \sum_{1}^{N} \delta_i(s_j) Y_i \big/ p_i \qquad (s_j \in S) \tag{2.15}$$

where $\delta_i(s_j)$ is an indicator random variable taking the value 1 if $Y_i \in s_j$ and 0 otherwise.

This estimator is known as the **Horvitz–Thompson estimator** (Horvitz and Thompson, 1952) and plays a central role in the unified treatment of finite population design and inference of Hedayat and Sinha (1991)—see also Thompson (1992) who stresses the applicability of the Horvitz–Thompson estimator to *any* sampling design (from simple random sampling to the most complex non-epsem scheme). Thompson emphasises the condition $i \in s_j$ by writing the estimator as

$$y_{HT} = \sum_{i=1}^{\nu} y_i/p_i$$

(after relabelling) where v is the number of *distinct* population members in the sample. In its general form, we find that

$$\boxed{E(y_{HT}) = Y_T}$$

so that y_{HT} is unbiased, and

$$\text{Var}(y_{HT}) = \sum_{i=1}^{N} Y_i^2 \left(\frac{1}{P_i} - 1 \right) + \sum_{i \neq j}^{N} \sum^{N} Y_i Y_j \left(\frac{P_{ij}}{P_i P_j} - 1 \right) \tag{2.16}$$

where we can estimate $\text{Var}(y_{HT})$ by

$$s^2(y_{HT}) = \sum_{i=1}^{v} \frac{y_i^2}{P_i} \left(\frac{1}{P_i} - 1 \right) + 2 \sum_{i=1}^{v} \sum_{j>i} \frac{y_i y_j}{P_{ij}} \left(\frac{P_{ij}}{P_i P_j} - 1 \right). \tag{2.17}$$

Approximate confidence intervals for Y_T (and $\bar{Y}$) can be obtained in the usual way employing the normal approximation.

A simplified form of (2.16) arises for the fixed-size design of sample size n (all $s_i \in S$ are of size n); see Hedayat and Sinha (1991, pp 47–48).

Obviously, the Horvitz–Thompson estimator becomes more and more efficient the closer we come to ensuring that $P_i \alpha Y_i$ i.e. the closer to pps sampling.

The variance estimate (2.17) can be difficult to calculate and can even yield negative values. A resolution to this problem—typically avoiding negative values but with some upward bias—is to be found in Brewer and Hanif (1983, p 68). Further illustration, derivation of results and extensions to the use of the Horvitz–Thompson estimator can be found in Thompson (1992, pp 49–55) and *in extenso* in Hedayat and Sinha (1991, pp 46–65 and later).

2.10 Estimating a proportion, *P*

Consider an engineering process in which a special component is produced for use in the assembly of a car. If some dimension, Y, is not within required tolerances, the component will not be able to be used. To estimate the *proportion*, P, of useful components in a large batch (that is, those within the required tolerances), a sr sample of size n is taken and a count is made of the number, r, which have satisfactory values of Y. The population of Y values is not in itself of interest, we would like to know merely the proportion P of such values which lie within the tolerance limits.

Rather than studying a *quantitative* measure, Y, in relation to whether it satisfies some criterion, we may sometimes be concerned *directly* with some *qualitative* attribute or characteristic: for example, with the proportion P of the inhabitants of Renton who live in rented accommodation. Again, a sr sample of size n gives an indication of P. If r out of the n live in rented accommodation we might estimate P by the sr *sample proportion*,

$$p = r/n.$$

As in the case of the estimation of the population total, we can again readily modify the results for estimation of the *population mean*, $\bar{Y}$, to describe the properties of the estimator p.

Suppose P represents the proportion of members of a finite population of size N who possess some characteristic A (for example, have acceptable tolerances, or live in rented accommodation). We define a variable X_i describing the ith member of the population so that

$X_i = 1$ if the member possesses characteristic A,

$\quad = 0$ otherwise.

Then

$$X_T = \sum_1^N X_i = R$$

is the number of members possessing the characteristic A. Consequently,

$$\bar{X} = \frac{1}{N}\sum_1^N X_i = R/N = P,$$

so that the proportion P is merely the *population mean* for the X-values.

Likewise, the sample proportion p is just the *mean* $\bar{x}$ for the *sample of X* values. In discussing the performance of p as an estimator of P we are *once again considering the use of a sr sample mean to estimate the corresponding population mean.*

The only essential difference arises from the simple structure of the population of X-values, *where only the values 0 or 1 can occur.* This implies a relationship between the population mean $\bar{X}$ (or P) and the population variance S^2, which takes the form

$$S^2 = \frac{1}{N-1}\sum_{i=1}^N (X_i - P)^2 = \frac{NP(1-P)}{N-1},$$

with a corresponding effect on the sampling behaviour of the estimator p. We will put $Q = 1 - P$. Then from the results in Section 2.2 we have

$$\boxed{E(p) = P}$$

so that p is unbiased for P and

$$\boxed{\mathrm{Var}(p) = (1 - f)S^2/n = \frac{(N-n)}{(N-1)}PQ/n.} \tag{2.18}$$

But if P is unknown, S^2 will not be known. We can estimate S^2 by the unbiased estimator

$$s^2 = \frac{1}{(n-1)}\sum_1^n (x_i - \bar{x})^2 = npq/(n-1),$$

where $q = 1 - p$. Thus an *unbiased estimator* of Var(p) is given by

$$s^2(p) = (1 - f)pq/(n - 1).$$

Note that this is *not* the sample analogue $(N - n)pq/n(N - 1)$ of (2.18) as we might intuitively think, although in practice the difference is unlikely to be important.

If the sampling fraction f is negligible, the estimator of Var (p) takes the simple form

$$s^2(p) = pq/(n - 1).$$

This holds, in particular, when we are sampling from an infinite population.

2.11   Confidence intervals for *P*

In sampling attributes or characteristics to estimate a proportion P, we know more about the sampling distribution of our estimator, p, than in the corresponding situations of estimating $\bar{Y}$ or Y_T.

Indeed, the exact distribution of p is known!

The number, r, of the sample members possessing the required attribute has a *hypergeometric distribution*, with probability function

$$p(r) = \frac{\binom{R}{r}\binom{N - R}{n - r}}{\binom{N}{n}} \quad \max(0, n - N + R) \leqq r \leqq \min(R, n).$$

So we could, in principle, make exact probability statements about r as a basis for constructing confidence intervals for P. We could try to use published tables or charts of cumulative probabilities for the hypergeometric distribution (particularly those especially designed for the purpose of making confidence statements about P). For example, Chung and DeLury (1950) give charts of 90%, 95%, and 99% confidence limits for P for $N = 500$, 2500, and 10 000 for various values of n and p. Tables of cumulative probabilities for the hypergeometric distribution, for more modest values of N (up to 20), are given in Owen (1962). But such sources provide only a limited aid; it is not possible to obtain very accurate results from the charts, whilst direct tabulations cover a relatively small range of values and can require complicated inverse interpolation.

Instead, in the modern spirit, we could calculate required probabilities and conduct the inverse interpolations by computer, and some statistical computer packages contain such facilities.

However, it would be an advantage to be able to avoid the tedious calculations involved in the use of the hypergeometric distribution in favour of a simpler approach.

One obvious possibility is to use the *binomial distribution* as an approximation to the hypergeometric distribution. If n is small relative to both R and $(N - R)$, the lack of replacement of sampled members of the population can be ignored and r has essentially a binomial distribution, $B(n, P)$. We could use this binomial distribution to construct confidence intervals for P. But again we encounter some similar computational difficulties and would need to resort to inverse interpolation or use of tables and charts or to published charts or tables of confidence limits *per se* (e.g. those given in Neave

1978, or Fisher and Yates, 1973); again we could use an appropriate statistical computer package for greater accuracy.

For immediate use, however, we will find it convenient to go one stage further and use the *normal approximation to the binomial distribution*. Thus we will effectively assume that

$$p \sim N(P, (1 - f)PQ/n) \tag{2.19}$$

as a basis for constructing approximate confidence intervals for P. Notice how this is not the immediate extension of the argument supporting the binomial distribution for p, since some account is taken of the 'lack of replacement' in incorporating the fpc in Var(p). In comparison with (2.18), we have also omitted a factor $N/(N - 1)$ from Var(p), which is justified for the sizes of population where the normal approximation will be used.

The normal approximation (2.19) will be reasonable, provided:

  (i) n is not too large relative to R or $(N - R)$,
  (ii) the smaller of nP and nQ is not too small; for example min $(nP, nQ) > 30$ should suffice. If P is in the region of ½, much smaller values of nP will be acceptable. Of course the values of nP and nQ will not be known and will need to be assessed through their unbiased estimators np and nq.

We can determine confidence intervals for P from (2.19) in the usual manner of applying the normal approximation to the binomial distribution. Thus, from (2.19), we have

$$\Pr\left[\frac{|P - p|}{\sqrt{(1 - f)PQ/n}} < z_\alpha \right] = 1 - \alpha,$$

so that an approximate $100(1 - \alpha)\%$ *two-sided confidence interval* for P is given as the region between the two roots of the quadratic equation

$$P^2[1 + z_\alpha^2(1 - f)/n] - P[2p + z_\alpha^2(1 - f)/n] + p^2 = 0.$$

This can be simplified even further if n is sufficiently large. Replacing Var (p) in (2.19) by its unbiased estimator $s^2(p)$ we obtain an approximate $100(1 - \alpha)\%$ two-sided confidence interval for P in the form

$$p \pm \left[z_\alpha \sqrt{(1 - f)pq/(n - 1)} \right].$$

We could also introduce the usual continuity correction to take account of the fact that we are approximating a discrete distribution by a continuous one. This is particularly relevant when n is near the limit for justification of the normal approximation; it will tend to correct the length of the interval which would otherwise be too short.

The circumstances under which we may proceed to these various stages of approximation are not easily described in a concise manner. Some indication has been given in the discussion above. The principal determinants are the values of n and of NP and NQ relative to N. Some further details, with numerical illustration, are given by Cochran (1977, Chapter 3) and Thompson (1992, Chapter 5).

2.12   Choice of sample size in estimating a proportion

Consider the effect of the form (2.18) for Var(p). Clearly Var(p) will be a maximum when $P = Q = \frac{1}{2}$, so that for a given sample size n we will be able to estimate P *least accurately* when it is in the region of $\frac{1}{2}$. This effect is more fully assessed by considering the value of $\sqrt{PQ}$ (reflecting the *standard error* of p). For $\frac{1}{4} < P < \frac{3}{4}$, $\sqrt{PQ}$ only varies over the range (0.433, 0.500), and little change occurs in the accuracy of the estimator p. P needs to be in the region of 0.07 (or 0.93) before the standard error is reduced to 50% of its maximum value.

Example 2.6

Suppose we wish to estimate P with a sr sample large enough for the standard error (SE) of the estimator p to be no more than 2%. How large a sample would be needed? This will depend on what we mean by 'no more than 2%'. Is this a statement about the *absolute* value of the standard error, or do we require the standard error to be no more than 2% of P. In the lattercase, we would be concerned with the *relative* value of the standard error: with the ratio of the standard error to the mean which is known as the *coefficient of variation*? The results will be quite different in the two cases!

(i) *For an absolute value*, we want

$$\mathrm{SE}(p) = \sqrt{PQ/n} \leq 0.02$$

(assuming that the population is large so that the fpc can be ignored).

(ii) *For a relative value*, we want

$$\mathrm{SE}(p)/P = \sqrt{Q/nP} \leq 0.02.$$

When P is large the required sample sizes are similar in the two cases (for $P = 0.98$ we have $n = 49$, $n = 52$, respectively, in cases (i) and (ii)). For small values of P the required sample sizes are markedly different (49 and 122 500, respectively in cases (i) and (ii), when $P = 0.02$).

Clearly, when we prescribe a *relative* value for the standard error of p, the required sample size inevitably increases consistently with decreasing values of P. It becomes unreasonably large for most practical purposes, in spite of the fact that the accuracy requirement (a 2% *coefficient of variation*) may not seem to be particularly stringent.

Yet this is likely to be the type of requirement we would impose in situations where interest centres on estimating by Np the total number of individuals, $R = NP$, in the population who possess the defining characteristic. To keep the sample size manageable when P is small, we may well have to be content with a coefficient of variation somewhat larger than the 2% considered above.

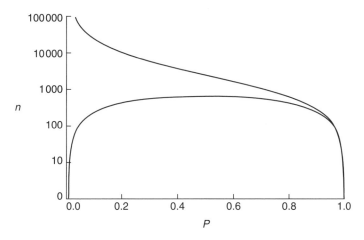

Fig. 2.3. Required sample sizes for '2% accuracy' in Example 2.6

The marked differences in the effects on required sample size of *absolute* and *relative* specifications of accuracy are shown (for the 2% specification of accuracy) in Figure 2.3.

The choice of a sample size to ensure certain limits on the standard error or coefficient of variation is of course equivalent to our earlier interest in achieving, with prescribed probability, a specified absolute, or proportional, accuracy for the estimator itself.

Thus to choose n to ensure that

$$\Pr(|\,p - P\,| > d) \lesssim \alpha, \tag{2.20}$$

$$\Pr(|\,p - P\,| > \xi P) \lesssim \alpha, \tag{2.21}$$

requires (using the normal approximation and ignoring the fpc) choice of n so that

(i) $SE(p) = \sqrt{PQ/n} \leqslant d/z_\alpha,$

or

(ii) $SE(p)/P = \sqrt{Q/nP} \leqslant \xi/z_\alpha,$

respectively.

In practical situations we must again recognise that the standard error, or the coefficient of variation, of p will not be known precisely since they depend on P, the quantity we are estimating.

However, one facility we now have which was not present when estimating $\bar{Y}$ or Y_T is that we can place *an upper bound* on the sample size required to achieve a required *absolute* accuracy in the estimation of P, *whatever the value that P happens to have*.

It's all relative

We can see this as follows. To satisfy (2.20) we need

$$n > PQz_\alpha^2/d^2.$$

But PQ has a maximum value of ¼, when $P = ½$. So that taking $n = z_\alpha^2/4d^2$ will certainly satisfy (2.20). Furthermore, this will not be too extravagant a policy over a quite wide range of values of P, say $(0.30 < P < 0.70)$. No similar facility is available if we want to ensure a certain *proportional* accuracy—see Figure 2.2 above.

Several loose ends remain to be tied up. It may be that the fpc cannot be ignored (since the sampling fraction may need to be sizeable to ensure the required accuracy).

Then again, we may be investigating a rather rare (or rather common) attribute in the population, so that the implication of assuming that $P = \frac{1}{2}$ in determining n to satisfy (2.20) will be to grossly *oversample* the population. Finally we may want to ensure a prescribed *proportional* accuracy.

Consider first of all the effect of retaining the fpc and using the exact form (2.18) for Var(p). To satisfy (2.20) and achieve a prescribed *absolute accuracy* we need (assuming the normal approximation to be justified)

$$n \geq N\left[1 + \frac{(N-1)}{PQ}\left(\frac{d}{z_\alpha}\right)^2\right]^{-1}$$

or, putting $(d/z_\alpha)^2 = V$,

$$N \geq \frac{PQ}{V}\left[1 + \frac{1}{N}\left(\frac{PQ}{V} - 1\right)\right]^{-1}. \tag{2.22}$$

So as a first approximation to the required sample size, we have

$$n_0 = \frac{PQ}{V},$$

which is just what was obtained above when we ignored the fpc

If n_0/N is *not* negligible, then we must use the more exact expression (2.22) to obtain

$$n = n_0\{1 + (n_0 - 1)/N\}^{-1}.$$

Correspondingly, from (2.21), for a required *relative accuracy*, we need

$$n \geq N\left[1 + (N-1)\frac{\xi^2 P}{z_\alpha^2 Q}\right]^{-1}$$

$$= \frac{Q}{P}\left(\frac{z_\alpha}{\xi}\right)^2\left\{1 + \frac{1}{N}\left[\frac{Q}{P}\left(\frac{z_\alpha}{\xi}\right)^2 - 1\right]\right\}^{-1}. \tag{2.23}$$

Hence, as a first approximation, we have

$$n_0 = \frac{Q}{P}\left(\frac{z_\alpha}{\xi}\right)^2$$

(*as obtained above for the required sample size when ignoring the fpc*). More accurately, when n_0/N is not negligible, we would take

$$n = n_0\{1 + (n_0 - 1)/N\}^{-1}.$$

In conclusion we must take account of the fact that P will not be known precisely (otherwise the survey would be pointless), so that (2.22) and (2.23) are not directly applicable. Again the methods (i), (ii), (iii), and (iv) of Section 2.6 must be considered as means of providing some 'advance estimate' of P for the purpose of determining the required sample size.

A pilot study will yield a preliminary estimate of P—this essentially combines methods (i) and (iv). The upper bound, provided by assuming that $P = \frac{1}{2}$ when *absolute* accuracy is of interest, also illustrates the use of method (iv). Again, and this is by no means uncommon, allied surveys (at an earlier time or on a related topic) may provide a reasonable idea of the value of P, as may published statistics on a wider front.

For example, suppose a large nationally based survey was conducted two years ago to investigate family possessions. It showed that 11% of families possessed digital television. If we decide to estimate the same measure on some particular local community we might choose to act as if P is somewhere in the region of 0.11 in order to determine a required sample size. Needless to say, we must ensure that the concept of the 'family' is a similar one in the two cases, and take note of any particular social structure in the local survey which may distinguish it from the national one (the two-year time separation is also highly relevant, of course, in a facility which is in the process of being introduced to the consumer).

Or again (cf. Section 2.6 above) we might use two-step sampling with the first-step estimate of P employed to guide the choice of sample size for the second-step (augmenting) sample. Specifically, if the first-step sample size is n_1 with estimate p_1, we should take a further $n - n_1$ observations to satisfy

$$n = \frac{p_1(1 - p_1)}{V} + \frac{3 - 8p_1(1 - p_1)}{p_1(1 - p_1)} + \frac{1 - 3p_1(1 - p_1)}{n_1 V}$$

to achieve a desired variance V, estimating P by $p + V(1 - 2p)/[p(1 - p)]$ where p is the usual full-sample estimate. Alternatively, we would take

$$n = \frac{1 - p_1}{C^2 p_1} + \frac{3}{p_1(1 - p_1)} + \frac{1}{C^2 n_1 p_1}$$

to achieve a desired coefficient of variation C in which case we estimate P by $\tilde{p} = p - Cp(1 - p)$ where p is the full-sample estimate. (See Cochran, 1977, Section 4.7, or Hedayat and Sinha, 1991, Section 4.6 for more detail.)

Example 2.7

Suppose, in the context of Example 2.3, that a somewhat liberal view is taken of 'casual holidays', recognising that the work is of such a nature that workers will feel the need to 'take the odd day off on the spur of the moment'. Up to 3 days in six months is regarded as reasonable; we want to estimate the proportion of workers taking more than 3 days off in the six months of study. From the figures given in Example 2.3 we find

$$p = 88/1000 = 0.088,$$

and obtain a 95% confidence interval for P, from Section 2.11, as

$0.071 < P < 0.105$

(Introducing the continuity correction yields $0.070 < P < 0.106$).
   The more accurate form obtained from solving the quadratic equation is

$0.072 < P < 0.107$.

   These results show some minor discrepancies. But these are hardly of any practical importance. Even ignoring the fpc gives a very similar interval: $0.0704 < P < 0.106$.

2.13 Further comments on estimating proportions

Sub-populations

In the industrial example on 'casual holidays' (Examples 2.3 and 2.7), it is likely that quite different patterns of behaviour exist for different groups of workers: administrative, technical, clerical, unskilled, etc. Different average numbers, and total numbers, of days 'casual holiday' may have been taken by the workers in the *sub-populations* for these different groups of workers. The population we have examined in Examples 2.3 and 2.7 is the combination of all the sub-populations; our estimates or inferences relate to this aggregate. But we may well wish to study the characteristics of the sub-populations separately. What group of workers takes the most, or least, number of days 'casual holiday'? What is the behaviour pattern for some particular group?

   This principle extends to estimating means, or totals.

   Thus if there are k sub-populations with means $\overline{Y}_i$, totals Y_{iT}, variances S_i^2, or proportions P_i ($i = 1, \ldots, k$), we may wish to estimate these characteristics separately. If this is our initial interest, we can take simple random samples from each sub-population at the outset and the methods described above and their corresponding properties will apply to each sub-population.

   But suppose we have taken a simple random sample from the aggregate population. How are we now to study the sub-populations? Subsequent assignment of observations to the sub-populations *does not yield* sr samples in these sub-populations — consequently the earlier results do not apply. The problem is that the sample sizes are not predetermined, but are themselves *random* quantities subject only to a constraint on their sum. More detailed study is now needed of how to analyse the data appropriately, in order to estimate the $\overline{Y}_i$, Y_{iT}, S_i^2, or P_i. Some of the results of such study are described by Cochran (1977, Chapter 2). See also Sections 4.6.3 and 4.6.4 below.

   Such structured populations present a complementary possibility. It may be that in recognising the different natures of the sub-populations, and sampling these sub-populations separately, we might be able to increase the efficiency of our study *of the aggregate population*. This is indeed true under certain circumstances, as we shall see later (Chapter 4) when we consider the ideas of *stratification*, and of *stratified sampling*, which constitutes an important refinement of the approach to survey sampling.

Multiple categories and multiple attributes

We have considered at some length the estimation of the proportion P of a population falling into some category. The classification of members of the population was a dichotomous one—each member was either in the category of interest, or not. Often a more complex classification exists, and population members must be assigned to one of *several* categories. For example, in an enquiry into social structure we may want to estimate the proportions of people in different social classes, where of course there are more than just two possibilities. Or again in the industrial example above, we might wish to estimate the proportions of workers in the different occupational groups.

Thus instead of a single proportion P (and the complementary proportion $Q = 1 - P$ *not* in the category of interest) we may have proportions $P_1, P_2, ..., P_k$, in k distinct categories ($k > 2$), that must be estimated from the appropriate assignment of the individuals in a simple random sample of size n. The numbers, $n_1, n_2, ..., n_k$, in the different categories now follow the natural extension of the hyper-geometric distribution of Section 2.11, but again if the n_i are small in relation to the population totals $N_1, N_2, ..., N_k$ we can reasonably approximate this distribution. In this case we would use the *multinomial distribution*

$$\frac{n!}{n_1! n_2! ...} P_1^{n_1} P_2^{n_2} ...$$

(with the constraints $\Sigma n_1 = n$, $\Sigma N_1 = N$ and $\Sigma P_1 = 1$) where previously we used the binomial distribution, as the basis for drawing inferences about $P_1, P_2, ..., P_k$.

Needless to say, if we are interested only in a single category i even when $k > 2$, then the results of Section 2.10 to 2.12 are directly applicable. We have merely to put $P = P_i$ and $Q = 1 - P_i$.

If we have several attributes $A, B, ...$ which are simultaneously of interest (each perhaps with more than two categories) we enter the more complicated realm of multivariate measures (and *complex surveys*: see Sections 5.5 and 6.6.1). Sometimes, but rarely, the attributes are independent and can be examined separately. Usually they are dependent and this simple prospect does not arise: consider, for example, the sex and the preference for different types of mobile phone, of the respondents to the industrial survey (a) of Section 1.2.

Proportion or ratio

Consider the Agriculture example of Section 1.2, concerning sales of fruit. Since fruit is a perishable product, some of it will not be sold because it is not of 'merchantisable quality'. Suppose we survey the weights of different fruit bought by greengrocers, and subsequently sold by them. It could be of interest to estimate the proportion of the fruit bought that is actually sold.

This is quite different however to our earlier idea of a proportion—where each observation is unambiguously classified as being in, or not in, some category with regard to the value of a *single* measured variable. For each sampling unit (e.g. a greengrocer) we *now* measure *two quantities*: purchases and sales (Y and X). We are interested in the population measure *of their ratio*. If we define a new variable $R = Y/X$, then estimation of the sample mean R is just a further example of estimating a population mean (but *not* a population proportion) from a sr sample of values of the derived single variable R, and the earlier methods apply. But we could proceed differently, seeking to estimate the population measure $\overline{Y}/\overline{X}$, perhaps using $\overline{y}/\overline{x}$. Now we are using the two quantities on each sampling unit. Our next topic for methodological study (in Chapter 3) will be the estimation of ratios which addresses this very problem.

2.14 Randomisation inference and model-based inference

In Section 1.6 we drew a distinction between *randomisation inference* and *model-based inference* as the basis for analysing and interpreting the sample outcomes of surveys.

In the case of **randomisation inference**, we take the view that the population is finite and immutable: a fixed set of N values of a variable Y. There is nothing random or indeterministic about the population values Y_i $(i = 1, 2, ..., N)$; it is just that we do not know all their values. So we take a sample from which to infer population characteristics. But statistical inference requires assumptions about variation and probability. In randomisation inference these arise *from the sampling scheme* (e.g. simple random sampling) and not from any probabilistic mechanism generating the population values (since the population is a fixed set of N values). Thus sampling behaviour of, say, the sample mean $\overline{x}$ is *a product of the sampling scheme*. It turns out to be unbiased, to have variance $(1 - f)S^2/n$, etc. because of the probabilistic and variational features of the different samples thrown up by random sampling. Such inferences are *descriptive*, or *enumerative*, describing only the finite set of values Y_i $(i = 1, 2, ..., N)$. This is the attitude and approach adopted throughout most of this book, reflecting the usual approach to finite population sampling and inference as the basis, for example, of *survey sampling*.

This is quite different from what happens in the statistical analysis of data from infinite populations, where it is assumed that each observation y entering a sample (and the population) does so in accord with a *probabilistic generating mechanism*: a *model* $\mathcal{P}$ which ascribes a probability (probability density) $p_\theta(y)$ to the occurrence of y in the sample. Here θ is a parameter which takes a value (often unknown) from some set Ω of possible values. Such a prospect *could* be considered for a finite population and there is indeed an extended approach to survey sample which does takes such a model into account.

We might, for example, assume that the Y_i $(i = 1, 2, ..., N)$ making up the finite population have arisen as N outcomes of the model $\mathcal{P}$. Thus we would have two components of variation now: one reflecting the generating mechanism of the model $\mathcal{P}$, the other reflecting the probabilities that the chosen sampling scheme would produce any particular sample of n observations from the population. So we would regard the sample as a random subset of the population (controlled by the sampling scheme) and the population as a random subset of an infinite population (controlled by the model $\mathcal{P}$): a sort of *super-population*. This approach to survey sampling is termed **model-based inference**.

An intermediate prospect of **model-assisted inference** is one where we wish to infer characteristics of a fixed or determined population (using the probabilistic properties of the sampling scheme) but we also take account of the fact that the population is likely to display certain features of smoothness, shape, relationship, etc. For example, we might expect a roughly symmetrically dispersed set of population values; or that the population mean and variance should be roughly equal (prompted by belief in a Poisson-type generating base); or that another variable X is highly correlated with Y over the finite population. Such assumptions are used (at least informally) in practice and appear in places throughout our treatment, such as in discussion of *choice of sample size* to achieve prescribed accuracy or in *ratio and regression estimators* or in *stratification* procedures.

How should we decide which of these approaches to adopt? Commonly, the pragmatic choice is to use the battery of techniques readily available under randomisation inference and most sample surveys are conducted on this basis. But it must be sensible to enquire whether our knowledge of a situation allows us to progress to model-assisted inference or even to full-blown model-based inference. Smith (2000) proposes a principle for deciding whether model-based methods are needed. Suppose you know all N population values. Do you know all you want to know about the situation? If so randomisation inference is appropriate. But if not, and uncertainty still exists, e.g. about what other population you might have encountered in this situation, then model assistance or model-based inference is called for.

The distinctions are discussed by Smith (1994, 2000) and by Rao (1999). Särndal *et al.* (1992) give details of how to conduct model-assisted survey sampling; see Valliant *et al.* (2000) for coverage of model-based inference. Barnett and Roberts (1993) describes an approach to testing for outliers in a finite population which leads to the need to consider a 'super-population' concept.

2.15 Exercises

2.1 Two independent sr samples of sizes 300 and 500 were chosen one after the other (without replacement) from a population of 2500 students in a non-residential College. Each student was asked the distance (in miles) from the College that he or she lived. The sample means and variances were

$$\bar{y}_1 = 2.98, \quad \bar{y}_2 = 3.42,$$
$$s_1^2 = 4.27, \quad s_2^2 = 3.68.$$

Calculate an approximate 99% confidence interval for the mean distance from the College that students live.

2.2 Consider the 'casual holidays' problem described in Examples 2.3 and 2.7. From company computer records it might be easy to obtain precise information on the number of workers who missed no workdays over the six months period of interest. Suppose that 49.82% of the workforce of 36 000 missed no work. An sr sample of 500 workers out of the remainder yielded the following results.

Days off	1	2	3	4	5	6	7	8	9	10
No. of workers	157	192	90	31	18	5	2	4	0	1

Estimate the total number of days 'casual holiday' taken over the six month period; and determine the approximate standard error of the estimator. Do the same calculations for the sr sample of size 1000 described in Example 2.3, and explain the discrepancies in the approximate standard errors in the two cases.

2.3 In a private library the books are kept on 170 shelves of similar size. The numbers of books on 16 shelves picked at random were found to be

28, 33, 25, 33, 31, 28, 22, 29, 30, 32, 26, 30, 21, 28, 25, 26.

Estimate the total number, Y_T, of books in the library, and calculate an approximate 95% confidence interval for Y_T.

Suppose the resulting estimate is not accurate enough. We want to be 95% sure that a sr sample estimate of Y_T is within 100 of the true value. How many shelves should be included in the sample?

2.4 An sr sample of size $2n$ is chosen from a finite population of size $N(N > 2n)$. The population mean and variance are $\bar{Y}$ and S^2, respectively. The sample is divided into two equal parts: the first n observations, and the second n observations. The sub-sample means are $\bar{y}_1$ and $\bar{y}_2$. Derive a simple unbiased estimator of S^2 based on n, $\bar{y}_1$, and $\bar{y}_2$.

2.5 A residential area has 5000 private houses. We want to estimate the proportions of houses with

(a) more than three persons living in them.
(b) more than one mobile phone owned by the occupants of the house.

The estimators are required to have standard errors not exceeding 0.01 and 0.02, respectively. From other surveys it would appear that the proportions, for (a) and (b), will lie in the ranges 0.35 to 0.55 and 0.80 to 0.90, respectively. The two proportions are to be estimated from a single sr sample. How large a sample is needed to meet the accuracy requirements?

3
Ratios: ratio and regression estimators

Sometimes we are interested simultaneously in two variables Y and X, particularly in the **ratio** of their population characteristics in the form $R = Y_T/X_T = \overline{Y}/\overline{X}$.

- Estimation of R (3.1) can be approached in various ways. Two particular estimators are the **sample average ratio**, r_1, and the **ratio of the sample averages**, r_2. Of these, r_1 is least tractable and useful although a modified form, the **Hartley–Ross estimator**, is of interest. For r_2 we have asymptotic unbiasedness and a simple form for its variance.
- For estimating population characteristics of Y in the face of concomitant information on X, the **ratio estimator** (3.2) of the population mean or total is defined in terms of r_2 and can be more efficient than the direct simple random sample mean or total, respectively, if X and Y are roughly proportional.
- More generally (3.3) the **linear regression estimator** also exploits the association between Y and X and can improve on the ratio estimator.
- Detailed comparisons are made (3.4) of the *ratio estimator* and the *linear regression estimator.*
- The effects of *pps sampling* are explored (3.5).

We have considered the problems associated with the estimation of a *single population characteristic*, based mainly on the probability sampling scheme of *simple random sampling*. Continuing with the same sampling scheme, we shall now broaden our enquiries a little with respect to the population characteristics of interest. Frequently (indeed predominantly) the aim of a sample survey is to seek information simultaneously on a *range* of different measures in the finite population we are studying. Both our practical interest and the cost and effort of conducting a survey demand that we should do so.

For instance, in the social affairs example (a) in Chapter 1, there will be a variety of *different* aspects of the response of 18 year olds to their recently acquired adult rights and responsibilities that will be of interest. The completed questionnaires will provide information on the range of such matters.

Initial contact with the respondents is the major cost and administration factor. To restrict attention to a single question ('Do you feel that the right to vote gives you an important say in the organisation of society?') is inefficient in cost terms. (Note also that this example is a leading question which can prejudice the reaction of the respondent!) To seek answers to, say, 20 questions involves little more trouble than obtaining the answer to one question; it provides far greater facility for assessing attitudinal factors and can yield wide-ranging information on the population for current or future use. Thus we are often confronted with multivariate data concerning a variety of measures in the population, represented by variables $Y, X, W, \ldots$.

Simultaneous estimation of population characteristics exploiting the correlation structure of the multivariate population is not a principal part of this introductory treatment. However, one simple extension of the univariate situation will be considered in detail in this chapter. This concerns the *bivariate* case where we simultaneously observe two variables, Y and X. We shall discuss two matters, distinct in aim but involving similar statistical considerations:

(i) how to estimate the ratio of two population characteristics, for example Y_T/X_T,
(ii) how simultaneous observation of Y and X, exploiting any association between these variables, can in certain circumstances assist in the efficient estimation of the characteristics of *one* of them, for example Y_T or $\bar{Y}$.

3.1 Estimating a ratio

In a variety of situations we may need to estimate a ratio of two population characteristics: the totals, or means, of two variables Y and X. We will be interested in the quantity

$$R = Y_T/X_T = \bar{Y}/\bar{X}, \tag{3.1}$$

which we will refer to as the **population ratio**.

This interest can arise in two ways. Either the ratio is of intrinsic interest in its own right. For example, we may wish to estimate the proportion of arable land given over to the growth of barley in some geographic region. To this end we might sample farms in the region and record their total acreage, and the acreage of barley crops. If these are X_i and Y_i for the different farms in the region, it is precisely $R = Y_T/X_T$ that we must estimate.

Alternatively, concern for the ratio R may arise from administrative convenience in the construction of a viable sampling scheme. Suppose we want to estimate the average annual income per head, or average number of cars per person, for adult persons living in a particular geographic region. We might envisage taking a simple random sample of adult individuals, noting their income or the numbers of cars they possess (predominantly 0 or 1) and using the sample mean in each case to estimate the corresponding population mean of interest. But it might not be easy to sample adults *individually* at random—ease of access to the population, and other quantities of interest in their own right, could favour the use of larger sampling units, say households. If this is the case, we become inevitably concerned with ratios, rather than means. The natural variables Y_i and X_i here are now household income and household size.

The average income per head is now best regarded as the ratio of total income Y_T to the total adult population size X_T, with both characteristics estimated from the sample of households. A similar situation arises for the car-ownership enquiry.

Note two features of this example: the use of groups of individuals (as sampling units) in the study of characteristics *per individual* (this relates to the idea of *clustering*, discussed in more detail later); also the simple nature of one of the variables in being discrete and taking only a few possible values (or even an indicator variable taking just the values 0 or 1). Both features commonly arise in ratio estimation (although the barley example shows that a simple discrete form for one of the variables is not inevitable).

Thus we wish to estimate the population ratio $R = Y_T/X_T$, on the basis of a simple random sample $(y_1, x_1), \ldots, (y_n, x_n)$ of the bivariate population measures (Y_i, X_i) $(i = 1, \ldots, N)$.

There are various possible approaches to estimation of R. Two immediately obvious ones are to use the **sample average ratio** or the **ratio of the sample averages**. Specifically, these are

$$r_1 = \frac{1}{n} \sum_{i=1}^{n} (y_i/x_i)$$

and

$$r_2 = \bar{y}/\bar{x} = y_T/x_T,$$

respectively. We will examine some of the sampling properties of r_1 and r_2.

We often find that Y- and X-values of practical interest are correlated; suppose Y is household expenditure on food and housing, whilst X is total household income. We must expect positive correlation between these two measures. Furthermore, it seems clear that the presence or absence of such correlations will affect the properties of the estimators r_1 and r_2. For example, with higher positive correlation the individual ratios Y_i/X_i will vary little compared with a corresponding uncorrelated situation (especially if S_Y^2 and S_X^2 are of similar order in the two situations) and we might expect this to reflect in the precision of the estimators. No correlation, or negative correlation, however would seem to be unpropitious (both in terms of likely practical interest and the properties of estimators of the population ratio).

(a)   Let us start by considering r_1

In spite of its intuitive appeal, r_1 is not widely used as an estimator of the population ratio R. It turns out to be biased and the bias and mean square error *can* be large relative to other estimators, particularly r_2. We can readily calculate the bias.

Consider the population of values $R_i = Y_i/X_i$. This has population mean $\bar{R}$ and variance S_R^2. Since r_1 is a sample mean from a sr sample, it has expected value $\bar{R}$ and variance

$$\left(1 - \frac{n}{N}\right) S_R^2/n.$$

But typically $\bar{R}$ is not the same as R, so we have

$$\text{bias}(r_1) = \bar{R} - R$$

$$= \frac{1}{N}\sum R_i - Y_T/X_T$$

$$= \frac{1}{N}\sum_1^N (Y_i/X_i) - \sum_1^N Y_i \Big/ \sum_1^N X_i$$

$$= -\frac{1}{X_T}\sum_1^N R_i(X_i - \bar{X}). \tag{3.2}$$

This features the covariance $S_{RX} = \sum_1^N R_i(X_i - \bar{X})/(N-1)$ between R and X.

Thus, noting that the mean square error is the sum of the variance and the square of the bias, we have

$$\text{MSE}(r_1) = \left(1 - \frac{n}{N}\right)S_R^2/n + (N-1)^2 S_{RX}^2/(X_T)^2.$$

We have the usual unbiased variance estimator $\sum_1^n (r_i - \bar{r})^2/(n-1)$ available for S_R^2 and we can readily obtain an unbiased estimator of the covariance S_{RX} in the form:

$$\sum_1^n r_i(x_i - \bar{x})/(n-1) = n(\bar{y} - \bar{r}\bar{x})/(n-1)$$

Hence (noting that $\bar{r} = r_1$) we can estimate the bias and the MSE of r_1 by means of

$$-(N-1)n(\bar{y} - r_1\bar{x})/[(n-1)X_T]$$

and

$$\left(1 - \frac{n}{N}\right)\sum_1^n (r_i - r_1)^2/n + (N-1)^2 n^2 (\bar{y} - r_1\bar{x})^2 [(n-1)^2 X_T^2]$$

respectively, *provided X_T is known* (a condition which will prove to recur throughout our study of ratio estimators, and which is not infrequently satisfied in practice).

So if X_T were known, we could actually correct r_1 for estimated bias obtaining a modified estimator

$$r_1' = r_1 + (N-1)n(\bar{y} - r_1\bar{x})/[(n-1)X_T]. \tag{3.3}$$

This is known as the **Hartley–Ross estimator**.

Another way in which we could eliminate the bias is to sample (*with replacement*) with probability proportional to the X_i values—rather than using sr sampling (see Section 2.9 and Section 3.5 below). Whilst this is not often possible in precise form, we can sometimes devise a sampling procedure which effectively achieves it: e.g. sampling from an electoral roll at random will tend to reflect different roads with probabilities proportional to their numbers, X_i, of registered voters.

Note the effect of such sampling. We now have

$$E(r_1) = E(Y/X) = \sum_{1}^{N}\left(\frac{Y_i}{X_i}\right)\frac{X_i}{X_T} = \frac{Y_T}{X_T} = R$$

and thus r_1 is *unbiased*, without the need for any modification. More details of this approach will be found in Section 5.3 below in the context of *cluster sampling*.

(b)   Let us now consider r_2

This estimator is more widely used. Although still biased (and with a skew distribution) in small samples, the bias and MSE tend to be lower than for r_1 (although precise claims in this respect are hard to justify and only rather limited empirical studies have been made). In large samples the bias becomes negligible and the distribution of r_1 tends to normality, thus enabling inferences to be drawn based on a normal distribution with appropriate variance, $\mathrm{Var}(r_2)$.

As with r_1, we have to deal with the complication that both numerator $\bar{y}$ and denominator $\bar{x}$ reflect random variation. Let us start again with the bias. We have

$$r_2 - R = (\bar{y} - R\bar{x})/\bar{x}$$

and taking a Taylor series expansion about the population mean $\bar{X}$ gives

$$r_2 - R = \frac{\bar{y} - R\bar{x}}{\bar{X}}\left(1 + \frac{\bar{x} - \bar{X}}{\bar{X}}\right)^{-1}$$

$$= \frac{\bar{y} - R\bar{x}}{\bar{X}}\left[1 - \frac{\bar{x} - \bar{X}}{\bar{X}} + \left(\frac{\bar{x} - \bar{X}}{\bar{X}}\right)^2 - \cdots\right].$$

As an approximation to the bias we can take the first two terms to obtain

$$E(r_2) - R = E\left(\frac{\bar{y} - R\bar{x}}{\bar{X}}\right) - \frac{1}{\bar{X}^2}E[(\bar{y} - R\bar{x})(\bar{x} - \bar{X})].$$

The leading term is zero since $E(\bar{y} - R\bar{x}) = \bar{Y} - R\bar{X} = 0$. Furthermore,

$$E[(\bar{y}(\bar{x} - \bar{X})] = \mathrm{Cov}(\bar{y}, \bar{x}) = \left(1 - \frac{n}{N}\right)S_{YX}/n = \left(1 - \frac{n}{N}\right)\rho_{YX}S_Y S_X/n,$$

where ρ_{YX} is the correlation between Y and X. Thus we find, as an approximation to the bias of r_2,

$$E(r_2) - R \doteqdot \frac{(1 - n/N)}{n\bar{X}^2}(RS_X^2 - \rho_{YX}S_Y S_X) \tag{3.4}$$

which can be small if ρ_{YX} is close in value to RS_X/S_Y. This is equivalent to saying that *the regression of Y on X is linear and through the origin*, or that *Y and X are roughly proportional* to each other (see below for more detail on this matter in a broader context).

Suppose we now consider large samples, utilising asymptotic results. We find the following approximate results:

$$E(r_2) = \bar{Y}/\bar{X} = Y_T/X_T$$

and

$$\text{Var}(r_2) = \frac{1-f}{n\bar{X}^2} \sum_{i=1}^{N} \frac{(Y_i - RX_i)^2}{N-1} \tag{3.5}$$

where f is the sampling fraction n/N.

These approximate results may be obtained by writing as before

$$r_2 - R = \bar{y}/\bar{x} - R = \frac{\bar{y} - R\bar{x}}{\bar{x}}$$

and replacing $\bar{x}$ in the denominator by $\bar{X}$, which should be reasonable in large samples in view of the *consistency* of the sr sample mean. We obtain

$$E(r_2 - R) = \frac{E(\bar{y}) - RE(\bar{x})}{\bar{X}} = \frac{\bar{Y} - R\bar{X}}{\bar{X}} = 0.$$

Also, on this approximation,

$$\text{Var}(r_2) = E[(r_2 - R)^2] = \frac{1}{\bar{X}^2} E[(\bar{y} - R\bar{x})^2].$$

But if we define

$$Z_i = Y_i - RX_i,$$

then $(\bar{y} - R\bar{x})$ is just the mean, $\bar{z}$, of a simple random sample of size n chosen from the population of Z_i values.

Since this derived population has *zero* mean, we have from (2.3)

$$\text{Var}(r_2) = (1 - f)S_z^2/(n\bar{X}^2),$$

where S_z^2 is the variance of the population of Z-values.

Thus,

$$\begin{aligned}
\text{Var}(r_2) &= \frac{1-f}{n\bar{X}^2} \sum_{i=1}^{N} Z_i^2/(N-1) \\
&= \frac{1-f}{n\bar{X}^2} \sum_{i=1}^{N} \frac{(Y_i - RX_i)^2}{N-1}, \text{ as given in (3.5).}
\end{aligned}$$

Equivalently, we can make use of standard results on the asymptotic form of the mean and variance of the ratio of two statistics. We have that

$$E\left(\frac{\bar{y}}{\bar{x}}\right) = \frac{E(\bar{y})}{E(\bar{x})} + 0(n^{-1}), \tag{3.6}$$

and

$$\mathrm{Var}\left(\frac{\bar{y}}{\bar{x}}\right) = \frac{\mathrm{Var}(\bar{y})}{[E(\bar{x})]^2} - \frac{2E(\bar{y})}{[E(\bar{x})]^3}\mathrm{Cov}(\bar{y}, \bar{x})$$
$$+ \frac{[E(\bar{y})]^2}{[E(\bar{x})]^4}\mathrm{Var}(\bar{x}) + 0(n^{-3/2}). \tag{3.7}$$

But

$$E(\bar{y}) = \bar{Y}, \quad E(\bar{x}) = \bar{X},$$
$$\mathrm{Var}(\bar{y}) = \frac{(1-f)}{n}S_Y^2, \quad \mathrm{Var}(\bar{x}) = \frac{(1-f)}{n}S_X^2,$$

so that from (3.6) we again demonstrate that r_2 is approximately unbiased in large samples. The bias in r_2 is seen to be of order n^{-1} as we have already seen in (3.4). From the first-order terms in (3.7) we can again obtain the approximation (3.5), using the fact that

$$\mathrm{Cov}(\bar{y}, \bar{x}) = \frac{1-f}{n}\mathrm{Cov}(Y, X) = \frac{(1-f)S_{YX}}{n},$$

where the covariance, S_{YX}, of the bivariate population of (Y_i, X_i) values is *defined* as

$$S_{YX} = \frac{1}{N-1}\sum_1^N (Y_i - \bar{Y})(X_i - \bar{X}). \tag{3.8}$$

The variance of r_2 is of order n^{-1} as we should expect, but we note that, in general, the approximation is correct to order $n^{-3/2}$. If the bivariate finite population manifests a roughly *normal* form, the error term in (3.5) is of order n^{-2} (in line with results for infinite bivariate normal populations), and the approximation for $\mathrm{Var}(r_2)$ is correspondingly more accurate.

But (3.7) also yields an alternative form for the approximation (3.5), namely

$$\mathrm{Var}(r_2) = \frac{1-f}{n\bar{X}^2}\{S_Y^2 - 2RS_{YX} + R^2S_X^2\}, \tag{3.9}$$

which explicitly includes the population covariance, S_{YX}.

Further details on the adequacy of the approximate form for the variance of r_2 are given in Cochran (1977, Chapter 6).

Once again we encounter the familiar problem that the variance of our estimator is expressed in terms of *population* characteristics, which will be unknown. Thus we will need to *estimate* Var(r_2) from our data, and it is usual to employ the direct sample analogue

$$s^2(r_2) = \frac{(1-f)}{n\bar{x}^2} \sum_{i=1}^{n} \frac{(y_i - rx_i)^2}{n-1}.$$ (3.10)

This differs from (3.5) by a term of order n^{-2}.

The sum of squares

$$\sum_{i=1}^{n} (y_i - r_2 x_i)^2$$

is most conveniently calculated as

$$\sum_{i=1}^{n} y_i^2 - 2r_2 \sum_{i=1}^{n} y_i x_i + r_2^2 \sum_{i=1}^{n} x_i^2,$$

echoing the alternative form (3.9). Note how it is unnecessary to correct the y_i and x_i values for their sample means, owing to compensation effects.

We remarked above that the exact distribution of r_2 is most complicated, but that it approaches normality in large samples (when sampling from large populations). Thus, for large samples, we can construct confidence intervals for R. If $s(r_2)$ is the sample estimate of the standard error of r_2, $[\text{Var}(r_2)]^{1/2}$, we have an approximate $100(1 - \alpha)\%$ *symmetric two-sided confidence interval* for R in the form

$$r_2 - z_\alpha s(r_2) < R < r_2 + z_\alpha s(r_2).$$

Example 3.1

A local newspaper conducts a survey of food costs by taking a simple random sample of 48 basic foodstuffs purchased in a large European supermarket in its area. Prices (in euros) for these items are recorded on two separate occasions, six months apart, the earlier ones being denoted x_i, the later y_i. The sample ratio, $r_2 = \bar{y}/\bar{x}$ gives an indication of the change in basic food prices over the six months period in the form of an estimate of the population ratio R of the mean prices of all foodstuffs on the two occasions.

The following results were obtained:

$$\bar{y} = 1.21, \quad \bar{x} = 1.14;$$
$$\sum y_i^2 = 92.71, \quad \sum x_i^2 = 84.32, \quad \sum y_i x_i = 85.64$$

Clearly the population size will be very large in relation to the sample size $n = 48$, so that we can ignore the fpc.

So the estimated ratio is 1.061: a 6.1% rise in prices over the six months in question. The approximate sampling variance of the estimator is

$$\frac{5.906}{48 \times 47 \times (1.14)^2} = (0.0449)^2$$

so that we have an approximate 95% confidence interval for R as $0.973 < R < 1.149$. Clearly any firm statement of an average food price *increase* in the supermarket would be unwise in the light of this wide range of values in the approximate interval, which reflects the small sample size in the survey. What other problems are there with such a survey?

3.2 Ratio estimator of a population total or mean

Suppose we wished to estimate the total UK local government expenditure in 2001 on some particular service (health or education, say; let us in fact choose the provision of recreational facilities for children). We might decide to do this by making specific enquiries on such expenditure in a simple random sample of the different county and metropolitan authorities throughout the country.

Clearly there is going to be large variation in the amounts spent on recreational facilities by the different authorities. This will reflect many factors, including their land area, number of inhabitants, available budgetary resources, and rural, urban, and industrial breakdowns. Most of the variation in provision stems of course from the differences in sheer size of the authority. We may have available a deal of information on such factors in the population. If so, it would surely be desirable to try to make use of our knowledge of the structure of the population to assess the representativeness of any random sample we may draw, or to guide the choice of the sample in an attempt to obtain a more efficient estimator.

We shall be considering at length in the next chapter the concept of *stratification* (the division of the population into non-overlapping groups, or *strata*, which represent its structure), and its use in the construction of what we hope will be better estimators than would be obtained from a simple random sample drawn from the unstratified population at large.

At this stage, however, we will consider an alternative means of exploiting known elements of the population structure in certain circumstances. This consists of the use of *ancillary quantitative information* to construct what is called the **ratio estimate of the population total** (or **mean**).

It seems reasonable, in the local authority example above, that expenditure on recreational facilities for children should change from one authority to another *roughly in proportion* to their number of inhabitants, or their total annual budgets. (Some slight anomalies may be observed in particularly small or large authorities, for largely rural or industrial ones, or for different age profiles of the children but the pattern of proportionality overall is likely to appear quite strong.)

Suppose Y_i denotes expenditure on recreational facilities for authority i, X_i denotes the number of inhabitants in the authority, and we sample both measures simultaneously and at random from the whole population, to obtain a sr sample of size n: $(y_1, x_1), \ldots, (y_n, x_n)$. The total number of inhabitants for the whole population, X_T, is likely to be known fairly accurately (for example, from census returns); we will also know the number, N, of local authorities in the population. But we could have *estimated X_T* from the sample by means of the estimator

$$x_T = N\bar{x},$$

where $\bar{x}$ is the sr sample mean. Similarly we could estimate the total expenditure Y_T (the characteristic of principal interest) by

$$y_T = N\bar{y}.$$

The estimate x_T has no interest in its own right (since we know X_T), *but it has the important advantage* that by comparing it with the population characteristic X_T we can informally assess the representativeness of the sample. If x_T is very much less than X_T, then in view of the rough proportionality of Y_i and X_i we would conclude that y_T is likely to underestimate Y_T; if x_T is too large, so is y_T likely to be. If the proportionality relationship were *exact* we would have

$$Y_i = RX_i \qquad (i = 1, \ldots, N), \tag{3.11}$$

where R is the population ratio, Y_T/X_T or $\bar{Y}/\bar{X}$, discussed in the previous section. Thus,

$$Y_T = RX_T,$$

and we could estimate Y_T by replacing R with the sample estimate, r_2, to obtain an estimate of the population total, Y_T, in the form

$$y_{TR} = r_2 X_T = \frac{X_T}{x_T} y_T. \tag{3.12}$$

N.B. We shall *not* consider using the sample average ratio r_1 to estimate R in this context, *but will restrict attention to the ratio of the sample averages, r_2, which henceforth will be denoted r* (i.e. suppressing the subscript).

The estimator y_{TR} is called the sr sample **ratio estimator of the population total**. Note that this achieves precisely the type of compensation we require for values of x_T which are fortuitously larger, or smaller, than the known value, X_T; it reduces, or increases, our estimate of Y_T accordingly.

The *exact* case is discussed simply to motivate the estimator (3.12)—if (3.11) held, one observation (y_1, x_1) would determine R precisely and hence $Y_T (= RX_T)$, so that 'estimation' of Y_T is trivial.

If the exact relationship (3.11) does not hold (it is hardly likely to do so in any practical situation), the same aim for compensation must still be sensible whenever there is 'rough proportionality' between the variable of interest, Y, and the ancillary (concomitant) variable, X. In such cases we can again use the ratio estimator (3.12).

If interest centres on the population mean $\bar{Y}$, rather than the total Y_T, then similar arguments support the use of the **ratio estimator of the population mean**,

$$\bar{y}_R = r\bar{X} = \frac{\bar{X}}{\bar{x}}\bar{y}. \tag{3.13}$$

Such ratio estimators have obvious intuitive appeal, but clearly we must attempt to identify the circumstances under which we obtain an important improvement in efficiency of estimation over the direct sr sample total, y_T, or mean, $\bar{y}$. This must involve a clearer statement of what is meant by 'rough proportionality'.

The sole sample statistic that is used is the *sample ratio*, r, whose properties have been discussed in some detail in the previous section (as r_2: the ratio of the sample averages).

Consider the estimator $\bar{y}_R$. From (3.6), $\bar{y}_R$ is *asymptotically unbiased*; in certain circumstances it is unbiased for all sample sizes, as we shall see shortly.

From (3.5) and (3.9) we see that the approximate variance of $\bar{y}_R$ (for large samples) is

$$\mathrm{Var}(\bar{y}_R) = \frac{1-f}{n}\sum_{i=1}^{N}\frac{(Y_i - RX_i)^2}{N-1}$$

$$= \frac{1-f}{n}(S_Y^2 - 2RS_{YX} + R^2S_X^2).$$

That is

$$\boxed{\mathrm{Var}(\bar{y}_R) = \frac{1-f}{n}(S_Y^2 - 2R\rho_{YX}S_YS_X + R^2S_X^2)} \tag{3.14}$$

where $\rho_{YX} = S_{YX}/S_YS_X$ is the *population correlation coefficient*. If the exact relationship (3.11) held, then $\mathrm{Var}(\bar{y}_R)$ would of course be zero. In practice this will not be so, but $\mathrm{Var}(\bar{y}_R)$ is clearly going to become smaller, the larger the (positive) correlation between Y and X in the population.

For estimating Y_T we have analogous results for y_{TR}. It is asymptotically unbiased, and has large sample variance

$$\frac{N^2(1-f)}{n}\sum_{i=1}^{N}\frac{(Y_i - RX_i)^2}{N-1}$$

or

$$\frac{N^2(1-f)}{n}(S_Y^2 - 2R\rho_{YX}S_YS_X + R^2S_X^2).$$

Again we will need to estimate $\mathrm{Var}(\bar{y}_R)$, or $\mathrm{Var}(y_{TR})$, from the sample, and the most convenient forms to use are

$$\frac{1-f}{n(n-1)}\left(\sum_{i=1}^{n}y_i^2 - 2r\sum_{i=1}^{n}y_ix_i + r^2\sum_{i=1}^{n}x_i^2\right),$$

and

$$\frac{(1-f)N^2}{n(n-1)}\left(\sum_{i=1}^{n} y_i^2 - 2r\sum_{i=1}^{n} y_i x_i + r^2\sum_{i=1}^{n} x_i^2\right),$$

respectively.

However, it has been suggested (Cochran, 1977, Chapter 6; Robinson, 1987; Rao, 1988) that it can be better to weight the sample estimate by the factor $(\bar{X}/\bar{x})^2$ thus using $\bar{X}^2 s^2(r_2)$ (see (3.10)) to compensate for a tendency for the sample estimate to be too large when $\bar{x}$ is much larger than $\bar{X}$ and vice versa.

Using the large sample forms for variances, and exploiting the asymptotic normality of the estimators, we can again obtain approximate confidence intervals for $\bar{Y}$ or Y_T in the usual way. A reasonable practical prescription for using the normal distribution, and the approximate form for the variance, is that the sample size should be at least 40, the sampling fraction no greater than 0.25, and that the ratios $S_Y/\bar{Y}$ and $S_X/\bar{X}$ should both be less than 0.10 $\sqrt{n}$. These latter quantities are known as the population *coefficients of variation* for the Y and X variables. We shall denote them by C_Y and C_X, respectively.

When the large sample results are inappropriate, the assessment of the properties of $\bar{y}_R$ and y_{TR}, and the construction of confidence intervals for $\bar{Y}$ and Y_T using ratio estimators, are most complicated. The exact results are incompletely known, and not very tractable. Some approximate results, which take account of the fact that the distribution of r frequently has positive skewness, are summarised by Cochran (1977, Chapter 6). See also Robinson (1987) and Rao (1988).

Let us return briefly to the question of the bias of the ratio estimators. The model (3.11) for the relationship between the Y_i and X_i values in the population was of little relevance apart from motivating the form of $\bar{y}_R$ or y_{TR}. We cannot expect it to hold in practice; indeed if it did, there would be no estimation problem! Relaxing the model slightly we might consider one for which

$$Y_i = RX_i + E_i, \tag{3.15}$$

with $\sum_x E_i = 0$, where $\sum_x$ denotes summation over all subscripts i for which $X_i = x$. (This is the finite population analogue of the classical linear regression model; see Section 2.14 on model-assisted inference)

In this case $\bar{Y} = R\bar{X}$ (as required from the definition of R), and in a sr sample of size n,

$$r = \frac{\bar{y}}{\bar{x}} = R + \frac{\bar{e}}{\bar{x}},$$

where $\bar{e}$ is the sample mean of the E values in the sample.

Clearly the conditional expectation

$$E(\bar{e} \mid x_1, \ldots, x_n) = 0$$

for this model, so that $E(r) = R$ and we conclude that *r is unbiased for all sample sizes.* But whilst more plausible than (3.11), the model (3.15) is again unlikely to be exactly satisfied by our finite population. At best we may find that the population is 'roughly' of this form, and we may consequently be less concerned than otherwise about possible bias in r, $\bar{y}_R$, or y_{TR}.

We have examined in some detail the properties of ratio estimators of a population mean or total, but a crucial question remains. *Under what circumstances, if any, should we use a ratio estimator in preference to a simple random sample mean or total?*

More effort is involved in obtaining $\bar{y}_R$ (or y_{TR}) than $\bar{y}$ (or y_T), albeit to only a slight degree since the major task is in designing and conducting the survey. Simultaneous measurement of two quantities, Y and X, often poses little more work than measurement of one alone. We can thus rule out differential costs as a major distinguishing factor in many, if not most, circumstances.

The prime consideration becomes *the precision of the estimation principle*; is $\bar{y}_R$ (or y_{TR}) more or less efficient than $\bar{y}$ (or y_T). The answer turns out to be that either possibility can arise, depending on the population correlation ρ_{YX} and the coefficients of variation C_Y and C_X.

We must identify the conditions under which $\mathrm{Var}(\bar{y}_R)$ is less than $\mathrm{Var}(\bar{y})$; that is where *the ratio estimator is the more efficient*. From (2.3) and (3.14) we see that

$$\mathrm{Var}(\bar{y}_R) < \mathrm{Var}(\bar{y})$$

if

$$R^2 S_X^2 < 2R\rho_{YX}S_Y S_X ,$$

that is, *if*

$$\boxed{\rho_{YX} > \frac{1}{2}\frac{C_X}{C_Y}} \tag{3.16}$$

So we see that a gain in efficiency is not in fact guaranteed; *we need the population correlation coefficient to be sufficiently large*. (In practice we would need to assess the criterion (3.16) from sample estimates of ρ_{YX}, C_Y, and C_X.)

But notice that however large ρ_{YX} turns out to be, we still *need not necessarily obtain a more efficient estimator* by using $\bar{y}_R$ (or y_{TR}). If

$$C_X > 2C_Y , \tag{3.17}$$

the ratio estimator $\bar{y}_R$ (or y_{TR}) cannot possibly be more efficient than $\bar{y}$ (or y_T) even with essentially perfect correlation between the Y and X values. Thus two factors are important for efficiency improvement from ratio estimators: the variability of the auxiliary variable X must not be substantially greater than that of Y (in the sense of (3.17)), and the correlation coefficient ρ_{YX} must be large and positive.

Nonetheless, *many practical situations are encountered where the appropriate conditions hold and ratio estimators offer substantial improvement over $\bar{y}$ or y_T.*

Reviewing the situation we need the following circumstances to hold:

(i) We must be able to observe simultaneously two variables Y and X which appear to be roughly proportional to each other (that is, which have high positive correlation).

(ii) The auxiliary variable X must not have a substantially greater coefficient of variation than Y.

(iii) The population mean $\bar{X}$, or total X_T, must be known exactly.

The 'rough proportionality' in (i) implies a more or less linear relationship through the origin. The fact that this is through the origin has not been formally considered above. Its major importance is a negative one. If Y and X were essentially linearly related, but the relationship did *not* pass through the origin, then we might be well advised to consider an alternative estimator known as the *regression estimator*. This estimator is discussed in the next section (see Section 3.3 below) where it is contrasted in terms of usefulness and efficiency with the ratio estimator and the simple random sample mean.

Example 3.2

Let us reconsider the *Global Games* team data given in Section 1.7. For this population, with Y denoting *weight* and X denoting *height*, we clearly have quite a strong positive association between the Y_i and X_i values, as shown in the scatter diagram, Figure 3.1. Using our knowledge of

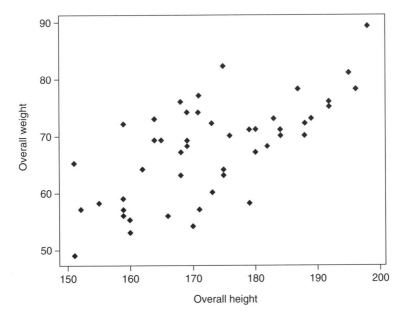

Fig. 3.1. Scatter diagram of weights (X) and heights (Y) in the *Global Games* team data

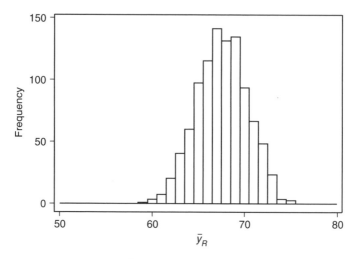

Fig. 3.2. Histogram of 1000 values of $\bar{y}_R$ from sr samples of size 5 in the *Global Games* team data

the whole population we can calculate various population characteristics of interest.

$$S_Y^2 = 73.95, \quad S_{YX} = 73.28, \quad S_X^2 = 155.67, \quad \rho_{YX} = 0.683,$$
$$R = \bar{Y}/\bar{X} = 0.390, \quad C_Y = 0.127, \quad C_X = 0.072.$$

The ratio C_X/C_Y of the coefficients of variation is about 0.57; the correlation coefficient very much in excess of half this value. Although the results above refer to large samples we should perhaps not be too surprised to find that the ratio estimator of $\bar{Y}$, based on a sr sample of size 5, greatly improves on the sr sample mean. To investigate this, 1000 such sr samples have been generated and $\bar{y}_R$ evaluated in each case. Figure 3.2 shows a histogram of the values obtained. By comparison with Figure 2.1, we see that $\bar{y}_R$ is far less disperse than $\bar{y}$; consequently more efficient. The mean and variance of the 1000 values of $\bar{y}_R$ are 67.54 and 7.46 respectively, in comparison with 67.53 and 13.61 for the 1000 values of $\bar{y}$. (The large sample approximation to $\mathrm{Var}(\bar{y}_R)$ is 7.28.)

3.3 Regression estimator of a population total or mean

Another form of estimator which aims at exploiting the relationship between some variable of interest, Y, and an auxiliary variable, X, in order to obtain greater precision in estimating $\bar{Y}$ or Y_T is the so-called **regression estimator**. This is a particularly useful estimator when (again) there is some degree of linearity in the relationship between the Y- and X-values in the population, but this relationship does

not necessarily pass through the origin. An *exact* relationship of this type would take
the form

$$Y_i = \bar{Y} + B(X_i - \bar{X}) \tag{3.18}$$

for all population values (Y_i, X_i) and some appropriate value of B. The regression
estimator, like the ratio estimator, is applied in situations where the value of $\bar{X}$ is
known. In the case of a population satisfying (3.18), we could clearly be able to
determine $\bar{Y}$ exactly from a single observation (y, x), since

$$\bar{Y} = y + B(\bar{X} - x).$$

But of course no such precise structure is likely to be encountered in real-life problems. What is possible, however, is that the Y- and X-values do seem on inspection to
vary in a way which reflects a degree of linearity, with relatively small superimposed
deviations about the linear relationship. Thus, for example, we might consider a
model in which

$$Y_i = \bar{Y} + B(X_i - \bar{X}) + E_i, \tag{3.19}$$

on the assumption that the E-values have zero population mean and bear no systematic relationship to the X-values, and in the belief that the population variance, S_E^2, of
the E_i will be rather small in relation to S_Y^2. The model (3.19) is then a useful representation of a population where the variation in Y-values may be attributed in part to
a linear dependence on the corresponding X-values, and in (perhaps lesser) part to
population vagaries unconnected with the X-values. If we assume (as suggested
above) that $S_{XE} = 0$, then we see how S_Y^2 has two components,

$$S_Y^2 = B^2 S_Y^2 + S_E^2,$$

and the population correlation coefficient is

$$\rho_{YX} = \frac{BS_X}{S_Y}, \tag{3.20}$$

since, from (3.19) and using $S_{XE} = 0$, we have

$$S_{YX} = B_x^2 S_x^2.$$

So,

$$S_E^2 = S_Y^2 (1 - \rho_{YX}^2),$$

and we note that the relative importance of the X- and E-values in accounting for the
variability in the Y-values depends on the value of ρ_{YX}^2. If the Y- and X-values are
highly correlated (in a positive or negative sense), the E-values make little contribution, and *vice versa*. This mirrors the characteristics of the classical linear regression
model in the infinite population case.

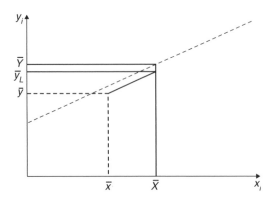

Fig. 3.3. Compensation effect of linear regression estimator $(B > 0; \bar{x} < \bar{X})$

Suppose now that we draw a sr sample $(y_1, x_1), \ldots, (y_n, x_n)$ and, knowing $\bar{X}$, want to estimate $\bar{Y}$. It seems sensible that we should take account of any linear relationship in the population, by using the estimator

$$\bar{y}_L = \bar{y} + B(\bar{X} - \bar{x}). \tag{3.21}$$

The estimator, $\bar{y}_L$, is called the **linear regression estimator** of $\bar{Y}$. Similarly $N\bar{y}_L$ is the linear regression estimator of the population total, Y_T.

Let us consider the justification for this estimator. In the extreme case (3.18) it is incontrovertible, for $\bar{y}_L \equiv \bar{Y}$. But no estimation problem exists in this case! For the more general model (3.19), $\bar{y}_L$ again seems a plausible choice. If $B > 0$, then if $\bar{x} < \bar{X}$ we would expect that $\bar{y} < \bar{Y}$ and (3.21) applies a compensation in precisely the manner we would wish, to take account of the linear dependence of Y on X. See Figure 3.3.

The same is true if $\bar{x} > \bar{X}$, or if $B < 0$.

Additional justification arises from considering the sampling behaviour of $\bar{y}_L$. It is clearly *consistent* in the finite population sense, since when $n = N$, $\bar{y}_L = \bar{Y}$. We see also that

$$E(\bar{y}_L) = E(\bar{y}) + B[\bar{X} - E(\bar{x})] = \bar{Y},$$

so that $\bar{y}_L$ is unbiased.

Furthermore,

$$\text{Var}(\bar{y}_L) = E[(\bar{y}_L - \bar{Y})^2] = E\{[(\bar{y} - \bar{Y}) - B(\bar{x} - \bar{X})]^2\}$$
$$= \frac{1-f}{n}(S_Y^2 - 2BS_{YX} + B^2 S_X^2).$$

So, from (3.20),

$$\text{Var}(\bar{y}_L) = \frac{1-f}{n} S_Y^2 (1 - \rho_{YX}^2). \tag{3.22}$$

Thus $\text{Var}(\bar{y}_L) \leqslant \text{Var}(\bar{y})$, and the efficiency of $\bar{y}_L$ relative to $\bar{y}$ increases with ρ_{YX}^2, that is, with increase in the correlation between the Y- and X-values.

Thus $\bar{y}_L$ has obvious advantages. It is unbiased for all sizes of sample and it cannot be less efficient than $\bar{y}$. Also we can obtain an *unbiased estimate* of $\text{Var}(\bar{y}_L)$ from the sample, in the form

$$\frac{1-f}{n}(s_Y^2 - 2Bs_{YX} + B^2 s_X^2),$$

where s_Y^2, s_{YX}, and s_X^2, are the familiar unbiased estimators of S_Y^2, S_{YX}, and S_X^2;

$$\text{e.g.} \quad s_{YX} = \frac{1}{n-1}\sum_{i=1}^{n}(y_i - \bar{y})(x_i - \bar{x}).$$

Example 3.3

Consider again the *Global Games* team data. If we take 1000 random samples of size 5, assuming that we know that $\bar{X} = 173.40$ and, for illustration, we put $B = 0.471$ (the population value), we find that the mean and variance of the 1000 $\bar{y}_L$ values are 67.60 and 7.32, respectively. So we again observe a *major efficiency gain over the simple random sample mean* of a similar order to that obtained for the *ratio estimator* of $\bar{Y}$ (see Example 3.2).

But for various reasons the results above are not as conclusive as they might appear. In practice, the exact value of the parameter B will of course be unknown. Furthermore, the model (3.19), with the additional assumption of zero correlation between the E_i and X_i, is unlikely to hold precisely; we cannot assess its appropriateness without studying the total population of (Y, X) values, in which case we would have no need for sampling or estimation. So how are we to relate these results to the practical situation?

We merely use our study of the behaviour of $\bar{y}_L$ in the special case above to *motivate* the consideration of a family of *general linear regression estimators* of the form

$$\bar{y}_L = \bar{y} + b(\bar{X} - \bar{x}) \tag{3.23}$$

(for different possible values of b) as a general principle of estimation. With no specific concern for the nature of any relationship between the Y_i and X_i, we might still ask whether an estimator of the form of (3.23) has any desirable properties as an estimator of $\bar{Y}$. The fact that it clearly does have such properties when (Y, X) satisfy (3.19) with $S_{XE} = 0$ and $b = B$ justifies such an enquiry.

We must consider two possibilities: either that b is given some pre-assigned value, or that we seek to estimate an appropriate value from the sample.

(a) Preassigned *b*

The sampling properties of $\bar{y}_L$ defined by (3.23) are already known to us from study of the special form (3.21).

We see that, *whatever the value of b*, $\bar{y}_L$ *is unbiased*, since

$$E(\bar{y}_L) = E(\bar{y}) + b[\bar{X} - E(\bar{x})] = \bar{Y}.$$

Also,

$$\mathrm{Var}(\bar{y}_L) = \frac{1-f}{n}(S_Y^2 - 2bS_{YX} + b^2 S_X^2) \tag{3.24}$$

with corresponding unbiased sample estimate

$$\frac{1-f}{n}(s_Y^2 - 2bs_{YX} + b^2 s_X^2).$$

An obvious question arises: if $\bar{y}_L$ is unbiased for all values of b, for what value of b does it have minimum variance? From (3.24) this must occur when

$$bS_X^2 - S_{YX} = 0$$

or

$$b = b_0 = S_{YX}/S_X^2 = \rho_{YX}\frac{S_Y}{S_X},$$

in which case $\mathrm{Var}(\bar{y}_L)$ takes the minimum value

$$\mathrm{Min\,Var}(\bar{y}_L) = \frac{1-f}{n}S_Y^2(1 - \rho_{YX}^2). \tag{3.25}$$

But this is just the same as (3.22), so we conclude that

$$\bar{y} + \rho_{YX}\frac{S_Y}{S_X}(\bar{X} - \bar{x})$$

is the most efficient estimator of $\bar{Y}$ of the form (3.23), *irrespective of any possible relationship between Y and X in the population*. If the model (3.19) happens to hold, then the optimum estimator is precisely the one, (3.21), which we considered for that model (and which had the practical appeal of applying an appropriate form of compensation to $\bar{y}$).

However b_0 will not be known in practice so that the optimal estimator is inaccessible. But we *may* be prepared to assign some specific value to b irrespective of the sample. This could happen if other studies of a similar nature have been carried out, and we feel fairly confident in transferring earlier knowledge to the present situation. Or again, the measures Y and X may be such that we anticipate a particular value for the slope of a linear relationship between them. In such situations we can use the sample approximation for (3.24) to assess the precision of our estimator, or compare it in efficiency with the sr sample mean, $\bar{y}$. We can also construct corresponding approximate confidence intervals for $\bar{Y}$ in the usual way, on the assumption of the normality of $\bar{y}$ and $\bar{x}$. (Appropriate conditions for this follow from the discussion in Section 2.5)

Furthermore, by considering the sample analogue of (3.25) we can informally assess how well our regression estimator compares in efficiency with the *best possible* estimator of the form (3.23). The relative efficiency is

$$(1 - \rho_{YX}^2)\left\{1 - 2b\rho_{YX}\left(\frac{S_X}{S_Y}\right) + b^2\left(\frac{S_X}{S_Y}\right)^2\right\}^{-1},$$

which, for large samples, may be reasonably estimated by substituting sample estimates of S_Y, S_X, and ρ_{YX}. We see that the proportional increase in variance due to using a non-optimal value for b is

$$\frac{\text{Var}(\bar{y}_L)}{\text{Min Var}(\bar{y}_L)} - 1 = \frac{\rho_{YX}^2}{1 - \rho_{YX}^2}\left(1 - \frac{b}{b_0}\right)^2. \tag{3.26}$$

where b_0 is the optimal value, $\rho_{YX}S_Y/S_X$. Again this is readily estimated from the sample.

(3.26) has important implications. Non-optimality of choice of the value of b can produce serious inefficiency in the regression estimator, relative to optimal choice. The relative inefficiency will be greatest in populations where Y and X are highly correlated. If the correlation is modest, choice of the value of b is less crucial, but then the potential gain over using $\bar{y}$ is very much less.

Example 3.4

Consider the expression (3.26) when $\rho_{YX} = 0.6, 0.8, 0.9, 0.95$ and $|1 - b/b_0| = 0.5, 0.2, 0.1$. The proportional increases in variance in these situations are:

$|1 - b/b_0|$	ρ_{YX}			
	0.6	0.8	0.9	0.95
0.5	0.141	0.444	1.066	2.314
0.2	0.023	0.071	0.171	0.370
0.1	0.006	0.018	0.043	0.093

Thus, for example when $\rho_{YX} = 0.95$ or 0.9 even modest discrepancies seriously affect the efficiency of the estimator.

(b) Estimated b

If we have no basis for preassigning a value to b, *as will most often be the case*, we might consider how we could use the sample data to suggest an appropriate value.

The optimal value of b, $\rho_{YX}S_Y/S_X$ expressed in terms of population characteristics, suggests that we might try using the corresponding sample expression

$$\tilde{b} = \frac{s_{YX}}{s_X^2} = \frac{\sum_{i=1}^{n}(y_i - \bar{y})(x_i - \bar{x})}{\sum_{i=1}^{n}(x_i - \bar{x})^2}. \tag{3.27}$$

(This has the same form as the least squares estimator of the classical linear regression coefficient for infinite populations.)

This is a reasonable procedure, at least in large samples. The regression estimator of the population mean now has the form

$$\bar{y}_L = \bar{y} + \tilde{b}(\bar{X} - \bar{x}). \tag{3.28}$$

Its distributional properties are difficult to determine precisely in view of the presence of the additional random variable $\tilde{b}$, which itself is a ratio of two statistics.

Large sample behaviour of (3.28) is more easily studied, and we base our investigation on the model (3.19). The extent to which a linear relationship is present will be reflected in the value of ρ_{YX} (as discussed above).

On the basis of this model we can show that the asymptotic forms of the expectation and variance of (3.28) are

$$\boxed{E(\bar{y}_L) = \bar{Y} + O(n^{-1})}$$

and

$$\boxed{\mathrm{Var}(\bar{y}_L) = \frac{1-f}{n}S_Y^2(1 - \rho_{YX}^2) + O(n^{-3/2})} \tag{3.29}$$

These results are comforting. Firstly we obtain an estimator which is unbiased in large samples. Secondly, and perhaps more importantly, we see that having to estimate b from the data is no disadvantage *in large samples*. We will obtain the optimum estimator: one with asymptotically as small a variance as is possible for this type of estimator. Thus, using $\tilde{b}$ must be preferable to assigning some arbitrary specific value to b, since at best this will yield an estimator with variance given by the leading term in (3.29), whilst if we are unfortunate in our choice of a value for b we may be faced with a far less efficient estimator.

But if all is well in large samples, the same cannot be claimed for small samples. The distinction between 'large' and 'small' samples in this respect requires a more detailed knowledge of how the mean and variance of $\bar{y} - \tilde{b}(\bar{X} - \bar{x})$ depend on sample size and population characteristics.

We shall not attempt to derive such results; indeed there is much that is not fully known about the sampling distribution of the regression estimator. We shall merely quote some of the known results. Further details on their derivation and implications can be found in the standard texts and research literature. Again, Cochran (1977, Chapter 7) is a useful source of information and references. See also Rao (1988).

On the question of bias, it happens that this becomes serious if there is marked evidence of a *quadratic* relationship between Y and X. Alternatively, it is aggravated by excess *kurtosis* in the set of X-values in the population. Correspondingly, the variance is seen to be most affected by the coefficient of *skewness* of the X-values, so that the large-sample approximation will be least accurate for highly skew populations (with respect to the X variable). A reasonable practical prescription for adopting the large-sample approximation for the variance of the regression estimator is that the sample size should be in excess of 50, and that the set of X-values in the population should not be greatly skew.

Again, in use we will need to *estimate* the variance

$$\frac{1-f}{n} S_Y^2 (1 - \rho_{YX}^2),$$

for which we will use

$$s^2(\bar{y}_L) = \frac{1-f}{n}(s_Y^2 - \tilde{b}s_{YX}).$$

Example 3.5

An environmental survey is to be conducted on the attitudes and preparedness of householders in a particular town, to a possible pollution hazard e.g. development of a new land-fill site for domestic and industrial waste. A list is available of the 14 800 householders in the town, and the survey is to be conducted by personal interviews which for a randomly chosen householder are estimated to cost £6.40. A total of £8000 is available to carry out the survey and setting-up costs are £1000. Major interest centres on estimating the population mean $\bar{Y}$ of some variable Y e.g. what extra local taxes an individual would accept to avoid the hazard. It is possible to observe, at an additional cost of £0.80 per individual, the value of a concomitant variable X (such as annual payments to charity) for which it is known that $\bar{X} = 42$.

A pilot study, and previous experience, combine to suggest approximate values for certain population characteristics as follows:

$$C_Y = 0.5, \quad C_X = 0.3, \quad \rho_{YX} = 0.60.$$

Should we use a ratio estimator, or sr sample mean, if processing costs in the two cases would be £250 and £200, respectively?

Firstly, consider estimating $\bar{Y}$ by $\bar{y}$ (the sr sample mean). We have, for sample size n,

$$SE(\bar{y}) = [(1 - n/14\ 800)]^{1/2} S_Y / \sqrt{n}$$

so

$$\text{SE}(\bar{y})/\bar{Y} = [(1 - n/14\ 800)]^{1/2}\, C_Y \big/ \sqrt{n}$$

But using the cost information

$1000 + 6.40n = 8000$, so that $n = 1094$.

Thus for the sr sample mean we achieve an estimator $(\bar{y})$ for which

$$C_{\bar{y}} = \text{SE}(\bar{y})/\bar{Y} = 0.0145$$

If we use the ratio estimator $\bar{y}_R = (\bar{y}/\bar{x})\bar{X}$ we have for sample size n',

$$\text{SE}(\bar{y}_R) = [(1 - n'/14\ 800)]^{1/2}(S_Y^2 - 2R\rho_{YX}S_Y S_X + R^2 S_X^2)^{1/2}\big/\sqrt{n'},$$

where $R = \bar{Y}/\bar{X}$. Hence

$$\text{SE}(\bar{y}_R)/\bar{Y} = [(1 - n'/14\ 800)]^{1/2}(C_Y^2 - 2\rho_{YX}C_Y C_X + C_X^2)^{1/2}\big/\sqrt{n'}.$$

From the cost information,

$1050 + 7.2n' = 8000$, so that $n' = 965$

and for the ratio estimator we find

$$\text{SE}(\bar{y}_R)/\bar{Y} = 0.0124.$$

Thus the ratio estimator is the better choice in spite of the extra costs and we need a random sample of size 965.

3.4 Comparison of ratio and regression estimators

We considered above (Section 3.2) the circumstances under which the *ratio estimator*, $\bar{y}_R$, of $\bar{Y}$ is more efficient than the sr *sample mean*, $\bar{y}$. This arises if

$$\rho_{YX} > \frac{RS_X}{2S_Y} \quad \text{(where } R = \bar{Y}/\bar{X}\text{).} \tag{3.30}$$

We saw that, for large samples,

$$\text{Var}(\bar{y}_R) \doteqdot \frac{1-f}{n}(S_Y^2 - 2R\rho_{YX}S_Y S_X + R^2 S_X^2). \tag{3.31}$$

Correspondingly, we observed that the *regression estimator* $\bar{y}_L$ (with b estimated from the data) has

$$\text{Var}(\bar{y}_L) \doteqdot \frac{1-f}{n}S_Y^2(1 - \rho_{YX}^2), \tag{3.32}$$

and thus asymptotically it *cannot be less efficient than* $\bar{y}$.

There remains the comparison of $\bar{y}_R$ and $\bar{y}_L$. We have

$$\text{Var}(\bar{y}_R) - \text{Var}(\bar{y}_L) \doteq \frac{1-f}{n}(R^2 S_X^2 - 2R\rho_{YX}S_Y S_X + \rho_{YX}^2 S_Y^2)$$

$$= \frac{1-f}{n}(RS_X - \rho_{YX}S_Y)^2,$$

(3.33)

and we therefore conclude that asymptotically *the regression estimator must be at least as efficient as the ratio estimator* under all circumstances. From (3.33) we see that the only situation in which the ratio estimator can have the same efficiency as the regression estimator is when

$$R = \rho_{YX}\frac{S_Y}{S_X}.$$

(3.34)

But we have seen that $\rho_{YX}S_Y/S_X$ is the optimum choice b_0 for the parameter b, in the sense that it minimises the variance of $\bar{y}+b(\bar{X}-\bar{x})$. We saw also that it is the inevitable value for the parameter B in the model (3.19). Thus $\bar{y}_R$ and $\bar{y}_L$ are equally efficient only if

$$R = b_0 = B.$$

Let us look a little more closely at the comparison of $\bar{y}, \bar{y}_R,$ and $\bar{y}_L$ and at the role of any formal model expressing an element of linearity (or proportionality) in the relationship between the Y and X values in the population.

Note immediately that *we do not need to make any explicit assumptions about a possible linear relationship between Y and X to derive the properties of* $\bar{y}, \bar{y}_R,$ *and* $\bar{y}_L$ *described above*. Merely defining $\bar{y}_R$ by (3.12) and $\bar{y}_L$ by (3.28), we find that (asymptotically, hence approximately in large samples) they are unbiased and have variances given by (3.31) and (3.32), respectively.

Thus $\bar{y}_L$ is always more efficient than $\bar{y}$, except in the isolated case where $\rho_{YX} = 0$, when they are equally efficient. Also, the relative efficiencies of $\bar{y}_R$ and $\bar{y}$ are governed by the value of ρ_{YX} with $\bar{y}_R$ being more efficient if (3.30) is satisfied (with the prerequisite that $C_X \leqslant 2C_Y$), otherwise less efficient.

Finally, $\bar{y}_L$ will always be asymptotically more efficient than $\bar{y}_R$, unless the special relationship (3.34) happens to hold in the population, in which case they are equally efficient. If (3.34) holds, then so of course does (3.30) and the conditions are satisfied for $\bar{y}_R$ and to be more efficient than $\bar{y}$.

What are important in this comparison are the *relative* values of $\bar{Y}$ and $\bar{X}$, and of S_Y^2 and S_X^2, and the value of the correlation coefficient, ρ_{YX}. We have no need to formulate any linear model to express the relationship between Y and X. However, we could choose to set up such a model, and this would serve two purposes. Firstly, it would provide a practical motivation for initially considering types of estimator of the form of $\bar{y}_R$ or $\bar{y}_L$, as we have already seen. Secondly, since ρ_{YX} is a measure of *linear* association, such a model might help to illustrate more tangibly the comparison of $\bar{y}_R$ and $\bar{y}_L$.

With no implied constraints on the bivariate finite population of values (Y_i, X_i) we can freely declare that

$$Y_i = \bar{Y} + k(X_i - \bar{X}) + E_i,$$

(3.35)

or that

$$Y_i = k'X_i + E'_i,$$

(3.36)

with

$$\sum_{i=1}^{N} E_i = \sum_{i=1}^{N} E'_i = 0.$$

It must of course be true that $k' = R$.

But with no additional assumptions, such models are sterile; in their unconstrained forms, (3.35 and 3.36), they imply no linearity of relationship, or proportionality between Y and X. However complex the pattern of values (Y_i, X_i), this can be accommodated by suitable values E_i (or E'_i) depending on the X_i. But if we demand that such dependence is not to be entertained, say by postulating that

$$\sum_{i=1}^{N} E_i(X_i - \bar{X}) = \sum_{i=1}^{N} E'_i(X_i - \bar{X}) = 0,$$

the models become much more structured. They now represent linearity, or proportionality, with superimposed 'deviations' (or 'errors') uncorrelated with the X values.

As we have seen at the beginning of the previous section, we must now have, in (3.35),

$$k = B = \rho_{YX} \frac{S_Y}{S_X}.$$

Also in (3.36) we find, on multiplying each side by $(X_i - \bar{X})$ and summing over the whole population,

$$k' = R = \rho_{YX} \frac{S_Y}{S_X}.$$

Thus if (3.36), with

$$\sum_{i=1}^{N} E'_i(X_i - \bar{X}) = 0,$$

happens to be an appropriate model for our population, it can be re-expressed as (3.35) with $E_i = E'_i$, and we have

$$R = k' = k = B.$$

But this is precisely the condition that needs to hold for $\bar{y}_R$ and $\bar{y}_L$ to be equally efficient.

Hence from the practical viewpoint the tangible justification for using $\bar{y}_R$ will rest on any indication in the data of a *linear relationship through the origin* between the values of Y and X, with no suggestion of correlation between the X_i and the deviations, E_i. So such a relationship does turn out to have a formal basis. The likely inferiority of $\bar{y}_R$ relative to $\bar{y}_L$ will be indicated by observing that the Y and X values do not seem to be roughly *proportional* (in a positive sense) to one another—or do not appear even to be roughly linearly related.

In conclusion, we can informally summarise the conditions which will support the use of ratio or regression estimators in the following way.

(i) *We are concerned with estimating $\bar{Y}$ (or Y_T) in a finite population where for each Y_i value there is a value X_i of some auxiliary variable.*

(ii) *Y and X can be simultaneously sampled and the population mean $\bar{X}$ is known precisely.*

(iii) (a) *If the data (or general knowledge) suggest some reasonable degree of linear relationship between the Y and X values then we can expect to obtain a useful gain in efficiency over $\bar{y}$ (or y_T) by using the regression estimator $\bar{y}_L$ (or $y_{TL} = N\bar{y}_L$).*

(iii) (b) *If the linear relationship has positive slope and appears to pass through the origin, we can expect a similar gain, for slightly less computational effort, by using the ratio estimator $\bar{y}_R$ (or y_{TR}). This saving in computation is the only possible advantage of $\bar{y}_R$ over $\bar{y}_L$, and is only worth exploiting if there is a clear indication of proportionality i.e. of a positive linear relationship through the origin.*

Note that in our formal discussion and comparison of these estimators the sole concern has been for achieving (asymptotically) unbiasedness and minimum variance. Whilst we shall not consider here any other criteria of choice, we should not disregard alternative prospects. Questions of the cost involved in achieving a certain precision are most relevant and may lead to a principle of balancing cost against precision. We touched on this in **Example 3.5** above. In our study (in the next chapter) of estimators for stratified populations, we shall have cause to consider cost optimality as a criterion of choice in addition to the idea of minimising the variance of unbiased estimators.

3.5 Ratio estimates and pps sampling

In Section 3.1 we contemplated the idea of estimating a *ratio* using *pps sampling*. The notion readily extends to a *ratio estimator* of a population mean (or total).

Suppose we take a sample $(y_1, x_1) \ldots (y_n, x_n)$ without replacement in such a way that the probability $p(y_j)$ of obtaining the observation y_j is proportional to x_j: say $p(y_j) = kx_j = x_j / \sum_1^N X_j$.

Further suppose that we know $\bar{X}$ and that (for illustrative purposes at this stage) Y_i is proportional to X_i, i.e. $Y_i = \beta X_i$.

We see that sampling with probability proportional to the size of the ancillary variable X is just the same as pps sampling for Y, since

$$p_j = p(y_j) = x_j \Big/ \sum_1^N X_i = y_j \Big/ \sum_1^N Y_i.$$

Now consider the estimator

$$\tilde{Y}_T = \frac{1}{n} \sum_1^n y_j/p_j = \frac{X_T}{n} \sum_1^n (y_i/x_i). \tag{3.37}$$

We have seen in Section 2.9 that $E(\tilde{Y}_T) = Y_T$ with variance zero in this pathological case. Thus we have ideal unbiased estimates of Y_T (or $\bar{Y}$) as $\tilde{Y}_T$ (or $\tilde{Y}_T/N$).

But (3.37) is just based on the sample average ratio r_1 (see Section 3.1) but with pps rather than sr sampling.

Although we will not in practice encounter strict proportionality, we can often come close to this condition and (3.37) remains a useful estimator. We have then the dual advantage in using (3.37) of obtaining a highly efficient estimator and of using a pps sampling scheme for the ancillary variable X which is likely to be easy to implement (whereas this might not be so for the principal variable, Y).

Examples come readily to mind:

Y	X
Yield of wheat	Acreage planted with wheat (sampled cartographically)
Productivity of a multi-outlet company	Number of employees in different outlets (sampled from an overall list of employees)
Household income	Size of household (sampled from Electoral Roll)

How are we to conduct pps sampling? There are various possibilities. If we have a list:

$$X_1, X_2, ..., X_N$$

(i) we can form the partial sums

$$X_1, X_1 + X_2, ..., \sum_1^N X_i (=X_T).$$

We then pick a random number Z in $(1, X_T)$ and choose X_j where

$$\sum_1^{j-1} X_i < Z \leqslant \sum_1^j X_i$$

But this is time consuming, and if we have such a list it can be easier to use the following approach.

(ii) *Lahiri's Method.* Suppose we know or can obtain an idea of the value of the largest X_j ($j = 1, 2, ..., N$). Call this X_{max}. We now pick at random and independently two numbers: one an integer in $(1, 2, ..., N)$ and the second a value in the range $(0, X_{max})$. Suppose we obtain j and x. If $x \leq X_j$ we take X_j as our observation; otherwise we reject the pair of numbers and try again. However, this method can be rather sensitive to errors in our assessment of the value of X_{max}. See Lahiri (1951) and Hedayat and Sinha (1991, pp 111 and 150).

Procedures based on physical structure are also often available, such as sampling proportional to geographic area merely by choosing a location at random on a map.

3.6 Exercises

3.1 Part of a coniferous forest contains 280 trees of the same species and of similar ages. A preliminary estimate is required of the total weight of timber that these trees will yield. A forestry expert claims to be able to make fairly accurate assessments of the yield from any tree merely by visual inspection, and makes such assessments for all 280 trees. He assesses the total yield as 439.5 tonnes. Subsequently, 25 trees picked at random are felled and their timber yields accurately determined. The actual yields, y_i, and corresponding assessed yields, x_i, provide the following summary results.

$$\sum_1^{25} y_i = 39.8, \quad \sum_1^{25} x_i = 41.4,$$

$$\sum_1^{25} y_i^2 = 69.08, \quad \sum_1^{25} y_i x_i = 70.64, \quad \sum_1^{25} x_i^2 = 73.47.$$

Estimate the total yield using either the ratio or regression estimator, whichever seems most appropriate. Compare the efficiencies of the ratio estimator, the regression estimator, and the estimator based on the sample of y_i values alone.

3.2 A field of wheat is divided into a large number of sampling units and the weights of grain (y_i) and grain plus straw (x_i) are recorded for a sr sample of size n. Additionally, the *total produce* (grain plus straw) for the whole field is weighed. Suppose that the coefficients of variation have the values

$$C_X = 1.0 \quad C_{XY} = 0.9 \quad C_Y = 1.1.$$

Calculate the gain in precision from estimating the total grain yield by means of a ratio estimator (using the x-values) rather than from the sample grain data alone.

Suppose it takes 15 minutes to weigh the grain on any unit and a further 2 minutes to weigh the straw: also 2 hours to weigh the total produce. How many units should be chosen for the ratio estimate to be more economical than the sr sample total in terms of the amount of time spent in carrying out the survey.

3.3 In studying lung function in a group of 560 workers in a textile factory an estimate was required of the mean value of some relevant measure Y. An sr sample of 10 workers was chosen and their Y values, y_i, determined by an appropriate test. A note was also made of their heights, x_i. The results were:

y_i	3.0	3.5	3.3	3.1	4.1	3.2	3.7	2.9	3.9	3.4
x_i (cm)	173	183	170	175	160	157	168	180	178	163

From routine medical records the average height for the group of 560 workers is known to be $\overline{X} = 173.2$ cm. Estimate $\overline{Y}$ from the data, and calculate an approximate standard error for your estimator.

4

Stratification and stratified simple random sampling

Sometimes a population falls naturally, or can be conveniently partitioned, into a set of sub-populations. Such a population is said to be *stratified*. If we take simple random samples from each stratum the overall sample is a *stratified simple random sample* (4.1).

- The *stratified sample mean* is readily defined, is unbiased for the population mean and has variance expressed in terms of the separate stratum weights and variances.
- Allocation of stratum sample sizes can be done in various ways; *proportional allocation* is particularly simple.
- A corresponding *stratified sample proportion* (4.2) can provide an attractive estimator of a population proportion.
- The *stratified sample mean* is more efficient than the simple random sample mean (4.3) if variation between stratum means is sufficiently large compared with within-strata variation.
- We can *optimise* this efficiency advantage by appropriate allocation of stratum sample sizes (4.4) and *choose an allocation* to yield a specified precision of estimation.
- The *extent of efficiency gain* of optimum over proportional allocation is easily determined (4.5).
- Difficulties of *not knowing stratum sizes and variances* (4.6.1), problems of *over-sampling of strata* (4.6.2), *interests in separate strata* or in *multiway stratification* (4.6.3), use of *post-hoc stratification* for bias reduction (4.6.4) and *optimum choice of stratum boundaries* (4.6.5) are all amenable to investigation.
- *Quota sampling* (4.7) is a form of multi-way stratified sampling commonly used in social and political sample surveys and opinion polls.
- *Ratio estimators* can be defined in a stratified population context (4.8) with interesting results.
- The complex conclusions of a study of stratified simple random sampling are drawn together in the concluding Section 4.9, followed (4.10) by some relevant *exercises*.

In our earlier studies of the precision with which we can estimate a population char-acteristic, $\bar{Y}$ say, from the analogous quantity $\bar{y}$ in a sr sample, we saw that a crucial factor was the variance S_Y^2 of the population. The larger the dispersion in the popula-tion, reflected in the value of S_Y^2, the less precise is the estimator $\bar{y}$ in the sense that its sampling variance is larger.

This is only to be expected; it makes sense intuitively, it is a feature of classical statistical methods for infinite populations, and it is reflected in the value of the sam-pling variance, $(1 - f)\,S_Y^2/n$, of $\bar{y}$ in finite populations.

Consider a simple numerical example. Suppose we have a finite population of 20 members in which Y takes values

$$6\ 3\ 4\ 4\ 5\ 3\ 6\ 2\ 3\ 2\ 2\ 6\ 5\ 3\ 5\ 2\ 4\ 6\ 4\ 5$$

Its mean is $\bar{Y} = 4$; its variance, $S_Y^2 = 40/19$. If we take a sr sample of size 5 and use the sr sample mean $\bar{y}$ to estimate $\bar{Y}$ we have

$$\text{Var}(\bar{y}) = 6/19 = 0.316.$$

Clearly we could obtain quite a range of different values for $\bar{y}$ in different samples; from 2.2 to 5.8. But notice the *structure* of the population in this case. It could be rearranged as

$$2\ 2\ 2\ 2\ 3\ 3\ 3\ 3\ 4\ 4\ 4\ 4\ 5\ 5\ 5\ 5\ 6\ 6\ 6\ 6$$

It consists of 5 groups, in each of which all 4 Y-values are the same. Suppose, for this rather special population, we had some mechanism by which we could choose just one member at random from *each* group to constitute our sample of size 5. We must then inevitably obtain, *on all occasions* that we draw a sample of size 5, the obser-vation's

$$2,\ 3,\ 4,\ 5,\ 6,$$

with sample mean 4. Thus our estimate has *no sampling fluctuation*; *its sampling variance is zero*, and the estimate is always equal to the population mean $\bar{Y}$ it is esti-mating. We seem to have found a perfect estimator! This extremely favourable situ-ation has arisen because we have been able to *remove all variability from within the defined groups* into which we have divided our population and from each of which we are taking a single observation at random to make up our required sample of size 5.

Can this artificial situation be encountered, at least approximately, in practice? If so, we should be able to reduce the sampling variance of an estimator below that encounted from sr sampling from the whole population.

Let us consider a more practical example. We want to conduct a survey to estimate the mean height of the schoolchildren in a small primary school with four classes, each of about 30 children, covering four different age groups. We decide to measure the heights of a sample of 20 children for this purpose. We might do this by picking 20 children at random from the playground during their brief mid-morning break. If all 120 children were present in the playground, we would hope to obtain a sr

Stratification

sample of size 20. But there would not be much time for this, and it is not obvious how we would seek to ensure that the sample is *random.*

It might be much easier to visit the four classes at lesson time and measure the heights of a sr sample of 5 children *from each class.* A stimulus for this approach is convenience or ease of sampling, but note the effect. The classes reflect natural groupings of the population and, because of the relationship between stature and age, *the heights in each such group are likely to be less variable than in the population at large.* This effect will be by no means as extreme as in the numerical example above, but we might still hope that the relative homogeneity of the groups will lead to *some* improvement in the efficiency of our estimate of the mean height, compared with a sr sample drawn from the total population.

If the population falls into natural groups, or can be so divided, it is called a **strati-fied population**. The examples suggest two possible advantages of such a structure:

(i) the stratification might be an aid to efficient estimation under appropriate conditions in the sense illustrated above,
(ii) the stratification may be particularly convenient in administrative terms making it easier to draw our sample.

Thus, it would appear that we may be able to estimate some population character-istic more efficiently by sampling from each group (*stratum*) separately than by sam-pling from the population at large, if it so happens that our variable Y shows *less variation within each stratum than in the total population*. This advantage cannot be guaranteed of course. Stratification chosen purely for administrative convenience (ease of access or of sampling) will not *necessarily* yield the required relative within-stratum homogeneity, but it will often do so.

Clearly it is worth considering the situation in more detail. In this chapter we will study how to estimate population characteristics in stratified populations, under what circumstances we can expect to obtain better estimators than those derived from a sr sample from the unstratified population, and the extent to which practical consid-erations may influence any potential efficiency gains from stratification. A sampling approach commonly used in opinion polls, market research, political surveys, etc. is **quota sampling**. This is a form of stratified sampling and we will consider it further in Section 4.7 below.

4.1 Stratified (simple) random sampling

Suppose we wish to estimate the mean, $\overline{Y}$, of the set of values $Y_1, Y_2, ..., Y_N$ in a finite population. We shall assume that the population is *stratified*, that is to say it has been partitioned into k non-overlapping groups, or *strata*, of sizes

$$N_1, N_2, ..., N_k \quad \left(\sum_{i=1}^{k} N_i = N \right)$$

with members

$$Y_{ij} \quad (i = 1, ..., k; \; j = 1, ..., N_i).$$

Thinking of each stratum as a sub-population we can carry over our earlier nota-tion, and denote the stratum means and variances by

$$\overline{Y}_1, \overline{Y}_2, ..., \overline{Y}_k \, ,$$

and

$$S_1^2, S_2^2, ..., S_k^2,$$

respectively.

The population mean and variance, $\bar{Y}$ and S^2, will of course have the forms

$$\bar{Y} = \frac{1}{N}\sum_{i=1}^{k} N_i \bar{Y}_i = \sum_{i=1}^{k} W_i \bar{Y}_i$$

(where $W_i = N_i/N$ is termed the *weight* of the ith stratum), and

$$S^2 = \frac{1}{N-1}\left\{\sum_{i=1}^{k}\sum_{j=1}^{N_i} (Y_{ij} - \bar{Y}_i + \bar{Y}_i - \bar{Y})^2\right\}$$

$$= \frac{1}{N-1}\left\{\sum_{i=1}^{k}(N_i - 1)S_i^2 + \sum_{i=1}^{k} N_i(\bar{Y}_i - \bar{Y})^2\right\}. \tag{4.1}$$

We shall assume that a sample of size n is chosen *by taking a sr sample of pre-determined size from each stratum*. The stratum sample sizes will be denoted $n_1, n_2, \ldots, n_k$ ($\sum_{i=1}^{k} n_i = n$). The sr sample from the ith stratum has members

$$y_{i1}, y_{i2}, \ldots, y_{in_i} \quad (i = 1, 2, \ldots, k),$$

and we denote the sample mean and variance in the ith stratum by

$$\bar{y}_i = \frac{1}{n_i}\sum_{j=1}^{n_i} y_{ij}$$

and

$$s_i^2 = \frac{1}{n_i - 1}\sum_{j=1}^{n_i} (y_{ij} - \bar{y}_i)^2.$$

In each stratum we have a *sampling fraction*, $f_i = (n_i/N_i)$ ($i = 1, 2, \ldots, k$).

Such a sampling procedure for the choice of a sample of total size n from the overall population is termed **stratified (simple) random sampling**.

At this stage we shall not discuss the basis of the stratification of the population— we shall just accept that the population is so stratified, and that we wish to estimate $\bar{Y}$ from the sample values y_{ij} yielded by stratified random sampling. Later, we shall consider how the stratification might be constrained by administrative considerations (sampling ease, costs, etc.) or, in contrast, how we can sometimes make use of any practical information we may possess about the population to effect a stratification likely to lead to particularly efficient estimation of $\bar{Y}$.

The estimator of $\bar{Y}$ commonly employed is the so-called **stratified sample mean**. This is defined as

$$\bar{y}_{st} = \sum_{i=1}^{k} W_i \bar{y}_i.$$

Note: this assumes that we know the stratum sizes N_i precisely, in order to determine the stratum weights $W_i = N_i/N$. We return to this in Section 4.6.1.

The stratified sample mean, $\bar{y}_{st}$, is not the same as the overall sample average

$$\bar{y}' = \frac{1}{n}\sum_{i=1}^{k} n_i \bar{y}_i$$

of the stratified random sample, except under the special circumstances where

$$\frac{n_i}{n} = \frac{N_i}{N}.$$

This would imply that the *sampling fractions* $f_i = n_i/N_i$ *are identical for all strata*; a special form of stratified sampling where the stratum sample sizes, n_i, are said to be chosen by **proportional allocation** (since the sample sizes are chosen to be *proportional* to the stratum sizes).

Such a principle can be time-saving with regard to the collection of the sample data. We shall see that it can have statistical advantages, but again it presupposes a knowledge of the stratum sizes, N_i. We assume throughout that this knowledge exists; otherwise the weights W_i will need to be *estimated* in some manner, with a resulting bias in the estimator $\bar{y}_{st}$ and loss of accuracy in the later derived results.

We must firstly consider the mean and variance of $\bar{y}_{st}$. We have

$$E(\bar{y}_{st}) = \sum_{i=1}^{k} W_i \bar{Y}_i = \bar{Y},$$

so that $\bar{y}_{st}$ is *unbiased*, in view of the inevitable unbiasedness of the stratum sample means, $\bar{y}_i$. Note that

$$E(\bar{y}') = \frac{1}{n}\sum_{i=1}^{k} n_i \bar{Y}_i,$$

so that the overall sample average $\bar{y}$ *will be unbiased only in the case of proportional allocation* (where $n_i/n = N_i/N$).

For the variance of $\bar{y}_{st}$ we have

$$\mathrm{Var}(\bar{y}_{st}) = \sum_{i=1}^{k} W_i^2 (1 - f_i) S_i^2 / n_i \qquad (4.2)$$

The proof is straightforward:

$$\mathrm{Var}(\bar{y}_{st}) = \sum_{i=1}^{k} W_i^2 \, \mathrm{Var}(\bar{y}_i)$$

provided (as is implied in the stratified random sampling procedure) that $\mathrm{Cov}(\bar{y}_i, \bar{y}_j) = 0$ if $i \neq j$, that is, the simple random sample means for different strata are uncorrelated. Thus we obtain (4.2), using (2.3) for $\mathrm{Var}(\bar{y}_i)$.

One further aspect of the behaviour of $\bar{y}'$ needs comment. Although with proportional allocation $\bar{y}_{st}$ and $\bar{y}'$ have the *same numerical value*, the overall sample average $\bar{y}'$ does *not* have the same variance as $\bar{y}$, the mean of a sr sample of size n from the whole population. The variance of $\bar{y}'$ takes on the appropriate special form of (4.2) for proportional allocation (see (4.4) below), whilst

$$\text{Var}(\bar{y}) = (1 - f)S^2/n,$$

with $f = n/N$. The reason for this discrepancy is the element of non-randomness of the stratified random sample that arises from the constraint that specific numbers, n_i, of members of the sample *must* be chosen from each distinct sub-population defined by the stratification.

Some special cases of (4.2) *should be considered.*

(a) *If the sampling fractions* $f_i = n_i/N_i$ *are negligible, we have*

$$\text{Var}(\bar{y}_{st}) = \sum_{i=1}^{k} W_i^2 S_i^2/n_i. \tag{4.3}$$

(b) *With proportional allocation,* $n_i = nW_i$, $f_i = f = n/N$, *we have*

$$\text{Var}(\bar{y}_{st}) = \frac{(1-f)}{n} \sum_{i=1}^{k} W_i S_i^2. \tag{4.4}$$

(c) *With proportional allocation, and constant within-stratum variances,* $S_i^2 = S_W^2$ ($i = 1, 2, ..., k$), *we have*

$$\text{Var}(\bar{y}_{st}) = \frac{(1-f)}{n} S_W^2. \tag{4.5}$$

Directly analogous results hold for the estimation of the population total, Y_T. Thus, from the stratified random sample we obtain an unbiased estimator $N\bar{y}_{st} = \sum_{i=1}^{k} N_i\bar{y}_i$, with variance $\sum_{i=1}^{k} N_i^2(1 - f_i)S_i^2/n_i$.

In real-life situations the stratum variances, S_i^2, will not be known. So if we wish to quote a standard error for the estimator $\bar{y}_{st}$, or to construct an approximate confidence interval for $\bar{Y}$, based on $\bar{y}_{st}$ (and involving its approximate normality in large samples), we will *need to estimate the S_i^2*. We have already considered this problem in the discussion (Section 2.4) of how to obtain an unbiased estimator of a population variance from a sr sample. The strata are just sub-populations and the sampled values in each stratum constitute a sr sample.

Thus, as

$$s_i^2 = \frac{1}{n_i - 1} \sum_{j=1}^{n_i} (y_{ij} - \bar{y}_i)^2 \quad (i = 1, 2, ..., k)$$

are (*unbiased*) estimators of the stratum variances S_i^2 ($i = 1, 2, ..., k$), we obtain *an unbiased estimator of* $\text{Var}(\bar{y}_{st})$ as

$$s^2(\bar{y}_{st}) = \sum_{i=1}^{k} W_i^2(1 - f_i)s_i^2/n_i$$

$$= \frac{1}{N^2} \sum_{i=1}^{k} N_i(N_i - n_i)s_i^2/n_i. \tag{4.6}$$

Naturally this requires that *all stratum sample sizes should be at least 2*, i.e. $n_i \geqslant 2$ ($i = 1, 2, ..., k$).

It is not entirely unreasonable that occasionally we may encounter some $n_i = 1$, typically when the population is highly variable and a very large number of strata need to be considered. Estimation of $\text{Var}(\bar{y}_{st})$ is still feasible in such situations: two ingenious methods for dealing with the extreme case where all $n_i = 1$ are described briefly by Cochran (1977, Section 5A.12).

In some situations the practical circumstances may suggest that *all stratum variances are equal*. If this is so, then it is desirable to combine the data from the different strata to obtain an overall, or 'pooled', unbiased estimator of the common variance s_W^2 in the form

$$s_W^2 = \frac{1}{N - k} \sum_{i=1}^{k} \sum_{j=1}^{n_i} (y_{ij} - \bar{y}_i)^2.$$

We can now estimate $\text{Var}(\bar{y}_{st})$ by

$$s^2(\bar{y}_{st}) = \frac{s_W^2}{N^2} \sum_{i=1}^{k} N_i(N_i - n_i)/n_i.$$

If in such a situation we can use proportional allocation in drawing the sample, we then have simply

$$s^2(\bar{y}_{st}) = \left(1 - \frac{n}{N}\right) s_W^2/n$$

as an unbiased estimator of $\text{Var}(\bar{y}_{st})$.

Approximate confidence intervals for $\bar{Y}$ (or Y_T) may again be constructed on the assumption of a normal sampling distribution for the estimator, $\bar{y}_{st}$. These will be given (for confidence level $1 - \alpha$) by

$$\bar{y}_{st} - z_\alpha s(\bar{y}_{st}) < \bar{Y} < \bar{y}_{st} + z_\alpha s(\bar{y}_{st})$$

or

$$N[\bar{y}_{st} - z_\alpha s(\bar{y}_{st})] < Y_T < N[\bar{y}_{st} + z_\alpha s(\bar{y}_{st})],$$

respectively.

As before, such approximations will only be reasonable if the conditions are satisfied (in terms of sample size, and so on) for $\bar{y}_{st}$ to have a distribution which is essentially normal, and for $s^2(\bar{y}_{st})$ to be close in value to $\text{Var}(\bar{y}_{st})$. The latter requirement is often the more stringent one, and replacement of z_α by a percentage point for an appropriate *t*-distribution is sometimes recommended.

But in stratified random sampling, the situation is far less clear-cut than in sr sampling from the overall population, and no general prescription for the construction of approximate confidence intervals is available. One difficulty is that although *total* sample size may be large, the sr samples within each stratum will frequently not be. Some work has to be done on what constitutes an 'appropriate' number of degrees of freedom to adopt when using a *t*-distribution in place of the normal distribution (see, for example, Cochran, 1977, p 96), but it can hardly be claimed to lead to any universally applicable policy.

4.2 Estimating a proportion

Suppose a proportion P of the overall stratified population possesses a particular attribute. The estimation of P from a stratified simple random sample presents no problems. Each stratum sample member can now be thought of as having a value x_{ij} ($i = 1, ..., k; j = 1, ..., n_i$) which is 1 or 0 depending on whether or not the sample individual possesses the defined attribute.

Clearly $P = \sum_{i=1}^{k}\sum_{j=1}^{N_i} X_{ij}/N = \bar{X}$ (the population mean X-value) and the stratified sample mean

$$\bar{x}_{st} = \sum_{i=1}^{k} W_i \bar{x}_i$$

will, on the argument above, provide an unbiased estimator of P. Thus we define an estimator of P in the form of a **stratified sample proportion**, p_{st}, by rewriting the above expression for $\bar{x}_{st}$ in the form,

$$p_{st} = \sum_{i=1}^{k} W_i p_i$$

where p_i is the sampled proportion (just $\bar{x}_i$) in the ith stratum.

If the actual proportions in the population strata are P_i ($i = 1, ..., k$), then the variance of p_{st} is just

$$\text{Var}(p_{st}) = \sum W_i^2 (N_i - n_i) P_i (1 - P_i) / [(N_i - 1)n_i]$$

which, ignoring terms in $1/n_i$, can be rewritten in the simpler form

$$\text{Var}(p_{st}) = \sum W_i^2 (1 - f_i) P_i (1 - P_i) / n_i \tag{4.7}$$

where f_i is the sampling fraction n_i/N_i in the ith stratum ($i = 1, ..., k$).

An unbiased estimator of the variance (4.7) is given by

$$\sum W_i^2 (1 - f_i) p_i (1 - p_i) / (n_i - 1) \quad \text{(see Section 2.10)}.$$

For proportional allocation (4.7) becomes

$$\text{Var}(P_{st}) = \frac{(1 - f)}{n} \sum_{i=1}^{k} W_i P_i (1 - P_i).$$

4.3 Comparing the simple random sample mean and the stratified sample mean

In the introduction to this chapter it was suggested that stratification of the population may on occasions increase the efficiency with which we can estimate population characteristics such as $\bar{Y}$ or Y_T. To examine this prospect let us compare $\bar{y}$ and $\bar{y}_{st}$ in the same situation. Both are unbiased estimators. Which is the more efficient, in the sense of having the smaller variance?

We know that

$$\text{Var}(\bar{y}) = (1 - f)S^2/n,$$

whilst $\text{Var}(\bar{y}_{st})$ is given by (4.2). To simplify the comparison we shall suppose that the stratified sample has been drawn *with proportional allocation*. Then, from (4.4),

$$\text{Var}(\bar{y}_{st}) = \frac{(1 - f)}{n} \sum_{i=1}^{k} \frac{N_i}{N} S_i^2$$

and

$$\text{Var}(\bar{y}) - \text{Var}(\bar{y}_{st}) = \frac{(1 - f)}{n}\left(S^2 - \frac{1}{N}\sum_{i=1}^{k} N_i S_i^2\right).$$

But, by (4.1),

$$S^2 = \frac{1}{N - 1}\left[\sum_{i=1}^{k}(N_i - 1)S_i^2 + \sum_{i=1}^{k} N_i(\bar{Y}_i - \bar{Y})^2\right].$$

Now if the stratum sizes, N_i, are large enough

$$\frac{N_i - 1}{N - 1} \doteq \frac{N_i}{N - 1} \tag{4.8}$$

and

$$S^2 \doteq \frac{1}{N}\left\{\sum_{i=1}^{k} N_i S_i^2 + \sum_{i=1}^{k} N_i(\bar{Y}_i - \bar{Y})^2\right\},$$

so that from (4.7)

$$\text{Var}(\bar{y}) - \text{Var}(\bar{y}_{st}) = \frac{(1 - f)}{nN}\sum_{i=1}^{k} N_i(\bar{Y}_i - \bar{Y})^2$$

$$= \frac{(1 - f)}{n}\sum_{i=1}^{k} W_i(\bar{Y}_i - \bar{Y})^2, \tag{4.9}$$

which is positive, except in the extreme case where the $\bar{Y}_i$ are all the same when it is zero. *Thus we conclude that the stratified sample mean will always be more efficient than the sr sample mean*, the more so the larger the variation in the stratum means.

But the assumption (4.8) *proves to be quite crucial.* Suppose the stratum sizes are not large enough for (4.8) to be a reasonable approximation. Then, using (4.1) for S^2, we obtain the more accurate expression

$$\text{Var}(\bar{y}) - \text{Var}(\bar{y}_{st}) = \frac{(1-f)}{n(N-1)}$$

$$\times \left\{ \sum_{i=1}^{k} N_i (\bar{Y}_i - \bar{Y})^2 - \frac{1}{N} \sum_{i=1}^{k} (N - N_i) S_i^2 \right\}, \qquad (4.10)$$

which need not necessarily be positive.

This more precise comparison shows that $\bar{y}_{st}$ is *not* necessarily more efficient than $\bar{y}$ under all circumstances. However, $\bar{y}_{st}$ *will be more efficient than* $\bar{y}$ if

$$\sum_{i=1}^{k} N_i (\bar{Y}_i - \bar{Y})^2 > \frac{1}{N} \sum_{i=1}^{k} (N - N_i) S_i^2 .$$

A further specialisation gives a more easily interpretable form for this condition. *Suppose all the strata have the same variance,* S_W^2. Then we require

$$\frac{1}{k-1} \sum_{i=1}^{k} N_i (\bar{Y}_i - \bar{Y})^2 > S_W^2. \qquad (4.11)$$

Thus, the stratified sample mean will be more efficient than the sr sample mean *if variation between the stratum means is sufficiently large compared with within-strata variation*; the greater this advantage the greater the efficiency of $\bar{y}_{st}$ relative to $\bar{y}$. (Note how this comparison mirrors the analysis of variance criterion in infinite normal samples for testing homogeneity of a set of means.)

Summarising the results of this section we can informally conclude that

- the higher the variability in stratum means, and
- the lower the accummulated variability of within-stratum Y-values over all the strata,

the greater is the potential gain from using the stratified sample mean $\bar{y}_{st}$ (rather than $\bar{y}$) for estimating $\bar{Y}$. The same will be true, of course, for the estimation of Y_T.

Example 4.1

It is interesting to see if we can illustrate the properties of $\bar{y}_{st}$, which are discussed above, by stratifying the population of weights given in the *Global Games* team data and studying corresponding stratified sample means. There are certain obvious, if rather artificial, methods of stratification in this situation, including

(i) by event (track and non-track: two strata),
(ii) by sex (male and female: two strata).

So in each case we have two strata of respective sizes 30 and 20. Suppose we take a stratified simple random sample of size 5 by proportional

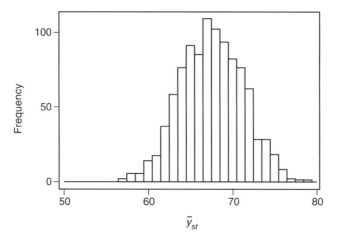

Fig. 4.1. Histogram of 1000 values of $\bar{y}_{st}$ (by event) for samples of size 5 in the *Global Games* team data

allocation in each case. For (i) this would yield three *track* athletes, out of 30, and two other (*field or multi-event*) athletes, out of the 20 of them in the team. For (ii) we would obtain three *male* and two *female* team members. It will be interesting to compare the sampling distributions of $\bar{y}_{st}$ (in each case) with the earlier derived empirical sampling distributions for $\bar{y}$ and for $\bar{y}_R$. Again 1000 random samples of size 5 were chosen, this time in accord with the two proportional allocation stratified sampling schemes. The sample averages and variances of the two forms of the stratified sample mean were as follows:

	Mean	Variance
Stratification by event	67.54	13.54 (13.03)
Stratification by sex	67.67	7.06 (6.94)

The variance figures in brackets are the exact values calculated from the population. So we see that stratification by event does not achieve an effective efficiency gain; the sampling variance is much the same as it was for $\bar{y}$. With such large differences between the statures of male and female athletes, we still obtain widely variable weights when we stratify by event. In contrast, stratification by sex effects a marked efficiency gain of similar order to that of $\bar{y}_R$ since the male weights and the female weights are likely to be less widespread than the weights over the whole population.

Figures 4.1 and 4.2 show histograms of the 1000 values of $\bar{y}_{st}$ for stratification by event and by sex, respectively.

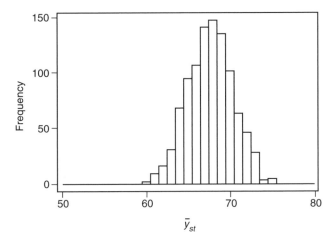

Fig. 4.2. Histogram of 1000 values of $\bar{y}_{st}$ (by sex) for samples of size 5 in the *Global Games* team data

4.4 Optimum choice of stratum sample sizes

So far it has been assumed that the total sample size n, and the stratum sample sizes n_i, have been *prescribed*. The study of the properties of $\bar{y}_{st}$ has assumed some particular allocation of the stratum sample sizes $n_1, n_2, \ldots, n_k$. As in all survey sampling schemes, we must remain aware of the need to achieve some required precision of estimation, and to do so either for minimum cost or (if possible) within some cost limitation imposed by the resources available for conducting the enquiry. Such factors are no less important in stratified sampling than for other sampling schemes. They may be even more important.

Often we have only limited choice of the basis of stratification. The major determinant is frequently an administrative one: that different methods need to be used to sample different sections of the population, or that natural (financial, geographic, social) divisions exist across which complete randomisation is clumsy, and unnecessary. For example, in a sociological study, lack of a common listing, differential problems of access and communication, desire for representative coverage of the population, and so on, could make it undesirable to attempt to sample at random the whole population (of hospital patients, prisoners, old-age pensioners, etc.).

Intrinsic interest in the sub-populations themselves also encourages stratification—a point we shall return to later (Section 4.6). Then again, in a national geographic survey, it is likely to be most convenient to sample different regions separately; costs of sampling can also vary from region to region (stratum to stratum) if only with respect to travelling expenses for survey workers.

So, again, we must consider the question of how to choose the sample size n to satisfy certain precision or cost requirements. Since different strata are likely to exhibit different degrees of variability, we must inevitably proceed beyond the choice of n to the allocation of the individual stratum sample sizes, n_i. Often we are interested in several population variables (e.g. household size, household income). What is a good stratum sample size allocation for one variable may be less reasonable for another. We will return to this dilemma (Section 4.6) but will proceed for the moment to discuss the allocation problem for one specific variable.

Further, to merely state our requirements in terms of the *variance* of some estimator will not be sufficient in general. *Different sampling costs for different strata* imply that we must attempt to take some account of cost factors in determining a desirable allocation of stratum sample sizes.

In any particular problem, local circumstances should enable a reasonably precise statement to be made of sampling costs in the different strata. Simple cost models have been proposed within which a large number of practical problems can be accommodated. The simplest form assumes that there is some overhead cost, c_0, of administering the survey, and that individual observations from the ith stratum each cost an amount c_i. Thus the total cost is

$$C = c_0 + \sum_{i=1}^{k} c_i n_i. \qquad (4.12)$$

This is the model which we shall adopt, although constant unit cost for observations sometimes exaggerates the true situation, and the replacement of $\sum_{i=1}^{k}(c_i n_i)$ by, say, $\sum_{i=1}^{k}(d_i \sqrt{n_i})$ may be better. This latter form is more reasonable, for example, when the major cost ingredient is for travel.

Suppose we adopt the cost model (4.12) and ask what allocation of stratum sample sizes, $n_1, n_2, \ldots, n_k$, should be adopted to

　(I)　minimise $\mathrm{Var}(\bar{y}_{st})$ for a given total cost C,
　(II)　minimise the total cost C, for a given value of $\mathrm{Var}(\bar{y}_{st})$.

We shall consider these cases separately to begin with.

I.　Minimum variance for fixed cost

We must choose $n_1, n_2, \ldots, n_k$ to minimise (4.2)

$$\mathrm{Var}(\bar{y}_{st}) = \sum_{i=1}^{k} W_i^2 S_i^2 / n_i - \frac{1}{N} \sum_{i=1}^{k} W_i S_i^2$$

subject to the constraint

$$\sum_{i=1}^{k} c_i n_i = C - c_0.$$

Introducing a Lagrangian multiplier*, λ, we will need

$$\frac{-W_i^2 S_i^2}{n_i^2} + \lambda c_i = 0 \quad (i = 1, \ldots, k),$$

or

$$n_i \sqrt{\lambda} = W_i S_i / \sqrt{c_i}.$$

---

*Readers unfamiliar with this technique may either accept the results (4.13) and (4.14) on trust, or may demonstrate them by the (lengthier) type of substitution argument used at the end of Section 2.2 (page 36).

Multiplying each side by c_i and summing over the strata, gives

$$\sum c_i n_i \sqrt{\lambda} = (C - c_0)\sqrt{\lambda} = \sum_{i=1}^{k} W_i S_i \sqrt{c_i}.$$

Thus

$$\sqrt{\lambda} = \sum_{i=1}^{k} W_i S_i \sqrt{c_i}/(C - c_0)$$

and the **optimum allocation for fixed total cost** is given by

$$n_i = \frac{(C - c_0)W_i S_i/\sqrt{c_i}}{\displaystyle\sum_{i=1}^{k} W_i S_i \sqrt{c_i}} \tag{4.13}$$

So the *total sample size* for overall cost C with optimum allocation will be

$$n = \frac{(C - c_0)\displaystyle\sum_{i=1}^{k} W_i S_i/\sqrt{c_i}}{\displaystyle\sum_{i=1}^{k} W_i S_i \sqrt{c_i}} \tag{4.14}$$

We see, then, that the stratum sample sizes will need to be proportional to the stratum sizes, proportional to the stratum standard deviations, and inversely proportional to the square root of the unit sampling costs in each stratum. (Large, highly variable strata with low unit sampling costs will lead to large samples relative to those from other strata.)

A special case of these results arises when the unit sampling costs c_i are the same in all strata, so that

$$C = c_0 + nc,$$

where c is the (constant) unit sampling cost.

The optimum allocation now needs

$$n_i = \frac{W_i S_i}{\displaystyle\sum_{i=1}^{k} W_i S_i} n \tag{4.15}$$

with

$$n = \frac{C - c_0}{c}. \tag{4.16}$$

Clearly the allocation (4.15) can equivalently be regarded as *optimum for fixed sample size, ignoring variation in costs of sampling from one stratum to another*, in the sense that, given n, it minimises Var($\bar{y}_{st}$).

The allocation (4.15) is given a special name: it is called **Neyman allocation**, after Jerzy Neyman who gave an early proof of its optimality (see Neyman, 1934; Tschuprow, 1923). The resulting minimum variance (for Neyman allocation; that is either for fixed sample size ignoring sampling costs, or with a cost limit and constant unit sampling costs) is obtained by substituting (4.15) in (4.2) to obtain

$$\mathrm{Var}_{\min}(\bar{y}_{st}) = \frac{1}{n}\left(\sum_{i=1}^{k} W_i S_i\right)^2 - \frac{1}{N}\sum_{i=1}^{k} W_i S_i^2, \tag{4.17}$$

the second term arising from the fpc In the cost-limited case, with constant unit sampling costs, n will need to be determined from (4.16).

An interesting question arises. If it were possible to choose the basis of stratification, how many strata should we have, and how should they be defined? We will return to this in Section 4.6.

II. Minimum cost for fixed variance

Suppose that, instead of putting a limit on total cost, we fix Var($\bar{y}_{st}$)—perhaps by imposing some precision requirement in the form that $\bar{y}_{st}$ should have a certain (high) probability of not differing from $\bar{Y}$ in absolute value by more than a specified amount (see **III** below).

We would now like to satisfy (for a prescribed value of V) a condition

$$\mathrm{Var}(\bar{y}_{st}) = V$$

for the minimum possible total cost. The appropriate allocation of stratum (and total) sample sizes is immediately determined from a reinterpretation of the results under **I**. We can see this informally from the following argument. We know that for any specified total cost, Var($\bar{y}_{st}$) is minimised when the n_i are chosen to be proportional to $W_i S_i / \sqrt{c_i}$. For given V there will be some total cost C for which this allocation yields V as the minimum variance. If C were made larger, then, in view of the explicit form of n_i given by (4.13), the minimised variance would *become less than* V; this we do not require. If C were smaller, by a similar argument we see that no allocation could restrict Var($\bar{y}_{st}$) to as small a value as V. Thus choice of n_i proportional to $W_i S_i / \sqrt{c_i}$ must also minimise total cost for a given value of Var($\bar{y}_{st}$).

Specifically we need

$$n_i = k W_i S_i / \sqrt{c_i},$$

where k must be chosen to ensure that

$$\mathrm{Var}(\bar{y}_{st}) = \sum_{i=1}^{k} W_i^2 S_i^2 / n_i - \frac{1}{N}\sum_{i=1}^{k} W_i S_i^2 = V.$$

Hence we must take

$$
n_i = \left\{ \frac{\sum_{i=1}^{k} W_i S_i \sqrt{c_i}}{V + \frac{1}{N} \sum_{i=1}^{k} W_i S_i^2} \right\} W_i S_i / \sqrt{c_i}, \tag{4.18}
$$

and the total sample size n will be

$$
n = \frac{\left(\sum_{i=1}^{k} W_i S_i \sqrt{c_i} \right) \left(\sum_{i=1}^{k} W_i S_i / \sqrt{c_i} \right)}{V + \frac{1}{N} \sum_{i=1}^{k} W_i S_i^2}.
$$

Again, if unit sampling costs are constant at some value c, we see that Neyman allocation (4.15) is optimum in the sense of *minimising total sample size* (since this is equivalent to minimising total cost) for a given Var$(\bar{y}_{st}) = V$. The resulting minimum total sample size will be

$$
n = \left(\sum_{i=1}^{k} W_i S_i \right)^2 \bigg/ \left(V + \frac{1}{N} \sum_{i=1}^{k} W_i S_i^2 \right).
$$

One more situation warrants investigation. Optimum allocation may not be feasible; suppose instead that we must operate with *prescribed sampling weights,* $w_i = n_i/n$, for the different strata. It is nonetheless useful to examine how to choose the total sample size to achieve a specified value for Var$(\bar{y}_{st})$.

III. Sample size needed to yield some specified Var($\bar{y}_{st}$) with given sampling weights

Suppose we want Var$(\bar{y}_{st}) = V$. For example, we might specify a margin of error, d, and acceptable probability of error, α, in the sense that we need

$$
\Pr \{ |\bar{y}_{st} - \bar{Y}| > d \} \leq \alpha.
$$

Proceeding as in Section 2.6, using an assumed normal distribution for $\bar{y}_{st}$, this would require

$$
V = (d/z_\alpha)^2.
$$

Equating Var$(\bar{y}_{st})$ (see (4.2)) to V gives

$$
n = \sum_{i=1}^{k} (W_i^2 S_i^2 / w_i) \bigg/ \left(V + \frac{1}{N} \sum_{i=1}^{k} W_i S_i^2 \right). \tag{4.19}
$$

Thus, as a first approximation to the required sample size, we have

$$
n_0 = \frac{1}{V} \left(\sum_{i=1}^{k} W_i^2 S_i^2 / w_i \right), \tag{4.20}
$$

or more accurately

$$n = n_0\left(1 + \frac{1}{NV}\sum_{i=1}^{k}W_iS_i^2\right)^{-1}. \tag{4.21}$$

In the special cases of *proportional allocation*, and of *Neyman allocation*, we have first approximations, n_0, and more accurate forms, n, given by

$$n_0 = \frac{1}{V}\sum_{i=1}^{k}W_iS_i^2, \quad n = n_0(1 + n_0/N)^{-1}, \tag{4.22}$$

and

$$n_0 = \frac{1}{V}\left(\sum_{i=1}^{k}W_iS_i\right)^2, \quad n = n_0\left(1 + \frac{1}{NV}\sum_{i=1}^{k}W_iS_i^2\right)^{-1}, \tag{4.23}$$

respectively.

It is important to ask how optimum allocation works out for estimating a population proportion, P (see Section 4.2). It turns out that the optimum allocations *for fixed sample size ignoring costs* (*Neyman allocation*) and *for fixed cost*, $C = c_0 + \sum_{i=1}^{k}(c_in_i)$, are

$$n_i = \frac{nW_i\sqrt{P_i(1 - P_i)}}{\displaystyle\sum_{i=1}^{k}W_i\sqrt{P_i(1 - P_i)}},$$

and

$$n_i = \frac{(C - c_0)W_i\sqrt{P_i(1 - P_i)/c_i}}{\displaystyle\sum_{i=1}^{k}W_i\sqrt{P_i(1 - P_i)c_i}},$$

respectively. The potential advantages of such optimum allocations over arbitrary or proportional allocation can be assessed from the appropriate forms of the results in Section 4.5.

In practice we will not know the P_i, and must use appropriate sample estimates. For example, we will estimate S_i^2 by the unbiased estimator $n_ip_i(1 - p_i)/(n_i - 1)$ (see Section 2.10).

4.5 Comparing proportional and optimum allocation

A crucial question to ask is to what extent optimum allocation of sample units to the different strata is more efficient than proportional allocation for estimating $\overline{Y}$.

Firstly, we note that proportional allocation is straightforward; it requires no knowledge of stratum variances or relative sampling costs. Secondly, such knowledge *is* required for optimum allocation in the sense of the previous section and it may not be available, or only known imprecisely. Its acquisition may require fairly detailed preliminary enquiries, or the acceptance of certain assumptions about the population structure that are difficult to justify. So proportional allocation has an importance in its own right.

Before considering (in Section 4.6 below) the implications for optimum allocation of an imprecise statement of sampling costs or stratum variances, we need to examine what *potential* gain can arise from optimum, rather than proportional, allocation.

We will consider just one case, the comparison of proportional allocation and Neyman allocation (optimum for constant unit sampling costs in different strata).

Denoting $\text{Var}(\bar{y}_{st})$ by V_P and V_N, for proportional and Neyman allocation, respectively, it must of course be true that $V_P \geqslant V_N$. More specifically, from (4.4) and (4.17), we have

$$V_P - V_N = \frac{1}{n}\sum_{i=1}^{k} W_i(S_i - \bar{S})^2, \tag{4.24}$$

where $\bar{S} = \sum_{i=1}^{k}(W_i S_i)$. Thus the extent of the potential gain from optimum (Neyman) allocation compared with proportional allocation depends on the *variability of the stratum variances*: the larger this is, the *greater the relative advantage of optimum allocation*.

Example 4.2

Consider once again the situation described in Example 2.5 where we wish to estimate the total Christmas card sales for a network of 243 stationery shops, by asking some of the shops to submit 'early returns' at the end of January. Suppose that for general accounting purposes the shops have been divided into three groups on the basis of their average annual turnover for all products over the period of the previous five years. The finite population is thus stratified with respect to annual turnover. Suppose that full March returns of the Christmas card sales *over recent past years* enable us to make fairly precise estimates of the three stratum variances for the current year. It is reasonable to expect that sampling costs will be higher for the larger shops and this turns out to be so.

Suppose that the strata sizes, variances, and sampling costs (in appropriate units) are as follows:

Average turnover (£'000)	N_i	S_i^2	c_i
Less than 50	146	0.16	2
Between 50 and 100	62	0.58	3
Greater than 100	35	0.31	4

We will assume that the aim is to estimate total sales for the current year, in such a way that we have a 95% chance of our estimate being within 10% of the true figure. We find that this requires (again assuming that total sales will be in the region of 420, and using the normal approximation) that we restrict the variance of our stratified sample estimator of total sales to about 460. Equivalently, we need $\text{Var}(\bar{y}_{st}) = V = (0.0882)^2$.

With proportional allocation we need sample stratum weights 0.601, 0.255 and 0.144, respectively. Now, following (4.19)–(4.21) with $w_i = W_i$,

$$\frac{1}{V}\sum_{i=1}^{k} W_i S_i^2 = 37.126,$$

and

$$37.126(1 + 37.126/234)^{-1} = 32.206,$$

so that allowing for the fact that sample sizes must be whole numbers, we will need to take 32 observations: with 19, 8 and 5 observations, respectively, drawn at random from the three strata.

Taking the calculations a stage further we might seek the optimum allocation. The sample weights now need to be about 0.527, 0.348, and 0.124, respectively. The required total sample size is again about 32 (31.71 in fact) and to meet the accuracy requirement with whole numbers of observations we need 17, 11, and 4 observations from the different strata.

What we should note from these results is the impressive reduction in the required sample size (from 62 to about 32) that arises from exploiting the stratification of the population. We would expect to obtain such an improvement from stratification in this situation, in view of the stratum variances being small relative to the population variance (which is of the order of 0.64), and in view of the fact that the stratum means will inevitably vary widely (being correlated with total turnover, our basis of stratification).

The reduction in total sample size for optimum, compared with proportional, allocation is negligible (zero in practice).

In examining the relative efficiency of $\bar{y}_{st}$ and $\bar{y}$ in Section 4.3, we assumed proportional allocation for the stratum sample sizes. It is interesting to see how the comparison differs for optimal (Neyman) allocation.

Denoting $\text{Var}(\bar{y})$ (for simple random sampling) by V and using the form (4.17) for V_N we have

$$V - V_N > 0$$

if

$$\left(\frac{1}{n} - \frac{1}{N}\right)S^2 - \bar{S}^2/n + \sum(W_i S_i^2)/N > 0.$$

Substituting for S^2 its form (4.2) in terms of stratum means and variances and assuming that stratum sample sizes are large enough for (4.8) to hold we find that $V - V_N > 0$ provided

$$\sum W_i(S_i - \bar{S})^2/n + \left(\frac{1}{n} - \frac{1}{N}\right)\sum W_i(\bar{Y}_i - \bar{Y})^2 > 0. \qquad (4.25)$$

Obviously both terms are non-negative and we *must* inevitably obtain an efficiency gain from stratified sr sampling with Neyman allocation (compared with simple random sampling) except in the limiting case where all stratum means are equal, and all stratum variances are equal. The efficiency gain will thus be the greater, the larger is the variability either in stratum means or in stratum variances.

It is useful to draw together the results of this section and of Section 4.3 to see what we have discovered about the *relative efficiencies* of the sr *sample mean* ($\bar{y}$) and the *stratified sample mean* under *proportional* allocation ($\bar{y}_{st(P)}$) and *Neyman allocation* ($\bar{y}_{st(N)}$). The following qualitative comparisons hold when stratum sizes are large enough for the approximation (4.8) to be reasonable.

$\bar{y}_{st(P)}$ versus $\bar{y}$	$\bar{y}_{st(P)}$ is at least as efficient as $\bar{y}$; the relative efficiency increases with *increase in the variability of the stratum means*
$\bar{y}_{st(N)}$ versus $\bar{y}$	$\bar{y}_{st(N)}$ is at least as efficient as $\bar{y}$; the relative efficiency increases with the *increase in the variability either of the stratum means or of the stratum variances*
$\bar{y}_{st(N)}$ versus $\bar{y}_{st(P)}$	$\bar{y}_{st(N)}$ is at least as efficient as $\bar{y}_{st(P)}$; the relative efficiency increases with *increase in the variability of the stratum variances*

There is a tidy logical development in these relative efficiency comparisons. However, we have already seen (in Section 4.3) that the situation becomes more complicated if (4.8) cannot be assumed.

For the comparison of $\bar{y}_{st(P)}$ and $\bar{y}$ we found that an efficiency gain required the variability of the stratum means to exceed a weighted combination of the stratum variances.

The comparison of $\bar{y}_{st(N)}$ and $\bar{y}_{st(P)}$ is unaffected by (4.8).

This leaves open the question of how crucial is (4.8) to the comparison of $\bar{y}_{st(N)}$ and $\bar{y}$. No simply expressed, or readily interpretable, intercomparison arises in this case and we shall not pursue it further, especially since (4.8) does not represent a major constraint and it is likely to be satisfied in most well-designed surveys.

4.6 Some practical considerations

We have now considered in some detail the effects, and possible advantages, of stratification of the population in relation to the estimation of $\bar{Y}$ and Y_T. The discussion assumed that the values of certain population characteristics—such as stratum sizes, N_i, and variances, S_i^2—are known. Little attention was given to real-life considerations in the choice of strata, or to the practical problems of determining stratum sample sizes when, as is likely, we do not have very accurate knowledge of the N_i or the S_i^2, or where more than one variable may be of interest. In this section we shall examine such matters a little more fully.

4.6.1   Effects of unknown N_i and S_i^2

The various results we have obtained for $\mathrm{Var}(\bar{y}_{st})$ under different circumstances, and for the choice of stratum sample sizes, have been expressed in terms of the N_i and S_i^2. In practice we may have no precise knowledge of the values of these quantities. At best we can estimate them from the survey data, or informally assign 'reasonable' values on the basis of previous experience of similar surveys. Any conclusions must reflect this lack of precise knowledge.

Even if we know the stratum sizes N_i, and adopt some prescribed allocation of stratum sample sizes n_i, then uncertainty of the values of the S_i^2 will mean that we cannot accurately assess the variance of the estimator $\bar{y}_{st}$ of $\bar{Y}$. The estimation of $\mathrm{Var}(\bar{y}_{st})$ from the sample data has been briefly discussed in Section 4.1.

If the N_i are also unknown, greater difficulties arise. The stratum weights $W_i = N_i/N$ are crucial ingredients in $\mathrm{Var}(\bar{y}_{st})$, and if we are to use a stratified sample mean we must estimate them in some way. It is possible that published data may help. For example, nationally based returns such as the Census (for human population factors), or other large-scale government department surveys (of agricultural, industrial, medical or educational factors), will contain a great deal of breakdown of information for the country as a whole. If it is reasonable to believe that some 'local' population *represents* the larger national environment, then we can adopt the national stratum weights in the local enquiry.

But such representativeness is far from inevitable—local populations can be notoriously idiosyncratic. The determination of stratum weights in this way *may* be perfectly reasonable—but it *can* be fraught with danger. The major safeguard lies in the experience, and shrewdness, of the investigator. Previous successes and failures must guide any decision to carry over 'global' stratum weights to the local problem.

Suppose we do not know the true stratum weights, W_i, and use instead estimated weights, U_i. The estimator of $\bar{Y}$ is now $\sum_{i=1}^{k} U_i \bar{y}_i$ which is clearly biased, with bias

$$\sum_{i=1}^{k}(U_i - W_i)\bar{Y}_i.$$

Note that this bias is fixed and independent of the sample size. The mean square error (MSE) of $\sum_{i=1}^{k} U_i \bar{y}_i$ is readily determined as

$$\mathrm{MSE} = \sum_{i=1}^{k} U_i^2(1 - f_i)S_i^2/n_i + \left[\sum_{i=1}^{k}(U_i - W_i)^2 \bar{Y}_i\right]^2 \tag{4.26}$$

and will need to be estimated from sample data. See Cochran (1977, Chapter 5A) and Sampford (1962, Chapter 6) for further discussion of related issues.

Thus, certain general conclusions can be reached concerning the likely effects of imprecise knowledge of the W_i. It will usually happen that $\bar{y}_{st}$ will be biased; its accuracy is best assessed by mean square error about $\bar{Y}$, rather than by $\mathrm{Var}(\bar{y}_{st})$. Furthermore, the bias does not tend to reduce with increase in sample size. The need to estimate the mean square error (necessary because the stratum variances will also be unknown) will introduce further imprecision.

Lack of knowledge of the N_i and S_i^2 (particularly the latter) becomes even more serious when choosing the allocation of stratum sample sizes. Such an allocation must be determined *before* we choose the main sample, so that we do not even have sample estimates of the S_i^2 to help us. Cochran (1977, Section 5A.1) considers this problem. Suppose (see (4.15)) we intend to use Neyman allocation with

$$n_i = nW_iS_i \Big/ \sum_{i=1}^{k} W_iS_i \tag{4.27}$$

to achieve minimum variance (4.17):

$$\text{Var}_{\min}(\bar{y}_{st}) = \frac{1}{n}\left(\sum_{i=1}^{k} W_iS_i\right)^2 - \frac{1}{N}\sum_{i=1}^{k} W_iS_i^2$$

but can only approximate the optimum allocation using sample sizes n_i' ($i = 1, 2, ..., k$) since the W_i and S_i^2 are not known. We would then actually attain a variance, from (4.2), of

$$\sum_{i=1}^{k} (W_i^2 S_i^2 / n_i') - \frac{1}{N}\sum_{i=1}^{k} W_iS_i^2.$$

Thus the loss incurred in terms of increase in variance due to failing to achieve the optimum (Neyman) allocation is

$$\sum_{i=1}^{k} \frac{W_i^2 S_i^2}{n_i'} - \frac{1}{n}\left(\sum_{i=1}^{k} W_iS_i\right)^2 = \left(\sum_{i=1}^{k} W_iS_i\right)^2 \sum_{i=1}^{k} \{(n_i' - n_i)^2 / (n^2 n_i')\}.$$

It can be shown that the proportionate increase in variance from departure from the Neyman allocation is correspondingly

$$\frac{1}{n}\sum_{i=1}^{k} \frac{(n_i' - n_i)^2}{n_i'} \tag{4.28}$$

(ignoring finite population corrections).

On some occasions, the potential gains from stratification are large enough to justify the expense of conducting some fairly detailed pilot survey *designed purely to estimate the stratum weights* as a preliminary to the main stratified sample survey. This approach is a further example of *double sampling*, or *two-phase sampling*, and makes it possible to take formal account of the sampling properties of the estimators of the W_i (implied by the method of drawing the pilot data) in assessing the properties of $\bar{y}_{st}$.

Double sampling of stratified populations is discussed by Cochran (1977; attributing basic theory to Neyman, 1938) and Thompson (1992). This approach can be useful not only when we do not know the N_i and S_i^2 but also in the not uncommon circumstances that units cannot be assigned to strata prior to selection. For example, we may sample individuals in a geographic region by random phone calls. Their sex, age, occupation, etc.—as bases for stratification—will not be known before they have been selected (even assuming they respond cooperatively to the ever more unpopular 'cold call' on their phone). Or again, we may collect biological specimens by quadrat but will not know the sample makeup before the quadrat contents are examined.

We can deal with unknown stratum membership of selected individuals (prior to selection) and unknown N_i and/or S_i^2 by *double (two-phase) sampling*. We first draw a sr sample of size n' in excess of the required sample size n we expect to need. We assign stratification as appropriate to all n' selected individuals, yielding n_i' individuals in stratum i, and estimate the N_i from them, using $w_i' = n_i'/n'$ as an unbiased estimate of $W_i = N_i/N$.

We then draw a stratified sr sample of n_i observations (typically at random from the n_i') from stratum i ($i = 1, 2, ..., k$) to yield a total sample size $n = \sum_{i=1}^{k} n_i$. Estimating the stratum weights w_i' from the first sample and the stratum means, as $\bar{y}_i$, from the second sample,

$$\bar{y}_{st}' = \sum_{i=1}^{k} w_i' \, \bar{y}_i$$

can be shown to be unbiased for $\bar{Y}$, with variance

$$\mathrm{Var}(\bar{y}_{st}') = S^2 \left(\frac{1}{n'} - \frac{1}{N} \right) + \sum_{i=1}^{k} \frac{W_i S_i^2}{n'} \left(\frac{1}{m_i} - 1 \right) \tag{4.29}$$

where $m_i = n_i/n_i'$ ($0 < m_i \leqslant 1$).

Often the attribution of stratum to first sample members can be achieved without actually measuring their *y*-values (possibly a costly process). If so the taking of an excessively large first sample may not be a particularly extravagant process.

We now have the questions of choice of n to achieve required precision or of the n_i for optimality. These need to take account of the respective costs of classifying first sample members in respect of their stratum occupancy and of measuring the *y*-values of second sample members. Results are readily obtained on these issues. Cochran (1977, Chapter 12) develops the arguments and presents the results—see also Thompson (1992, pp 143–144).

4.6.2 Over-sampling of strata

When applying the formulae (4.13) or (4.18) to determine optimum stratum sample sizes, it is not impossible that some of the resulting n_i may *exceed* the corresponding stratum sizes N_i. This is particularly likely to happen if the sampling fraction is large and if stratum variances differ widely. Obviously we cannot take more observations than there are members in a stratum, and the optimum allocation cannot be attained in the usual way. The common practice is to *fully sample* any strata for which the optimum n_i exceed N_i—that is, to take all members of those strata. For these strata $\bar{Y}_i$ and S_i^2 will be determined exactly. We can then reinterpret the sampling problem in relation to the reduced population—of strata not fully sampled. Thus we will take a stratified simple random sample from the reduced population with its smaller number of strata, using the expressions and principles above for forms of variances, and their estimators, and of choice of allocation schemes. If results are required for the overall population we will need to appropriately combine the deterministic (complete) information on some strata with the estimates (and their uncertainties) from other strata. The variance of the resulting $\bar{y}_{st}$ cannot now be obtained from the results for optimum allocation. We must take care to use the appropriate form of the general expression (4.2) (or its estimator (4.6)), recognising that certain sampling fractions f_i will be unity and hence that the corresponding terms in $\mathrm{Var}(\bar{y}_{st})$ will not contribute.

4.6.3 Sub-populations; multi-way stratification; several variables

In a sample survey we often wish to estimate characteristics of **sub-populations** (or '**domains of study**') as well as characteristics of the whole population. For example, in a survey of apple prices from different suppliers from a range of geographic regions we might be interested in average prices for each variety of apple over the different regions, or for each region over the different varieties of apple.

A stratified random sample will reflect both aspects but with different properties *depending on the criterion of stratification*; whether we have stratified by region or variety of apple or both, or neither. Consider the two main distinctions: where sub-populations of interest are the same as the strata, or where they are not.

(i) *Suppose the sub-populations are the same as the strata.* The samples in each stratum are just simple random samples of predetermined size. So we need only use the results of simple random sampling to estimate sub-population characteristics directly and to reflect the properties of the estimates. For example, if we stratify by variety of apple we can readily estimate means for the different varieties and assess the properties of the estimates. We recall that stratification is most advantageous for estimating overall population characteristics when between-stratum means differ widely and within-stratum variances are as small as possible. In this sense stratification by variety of apple might be better than stratification by region (although it may not be as convenient administratively).

(ii) *Suppose sub-populations do not coincide with the strata.* We might, for example, still be interested in means for different varieties of apple even if our stratification is by region. The sample might be made up of simple random samples from each region with the price of just one variety of apple (possibly chosen at random) from each sampled supplier. The sub-population sample for a variety of apple now consists of some observations from each stratum (region). *It is no longer a simple random sample*: the sample size is now a random quantity. Appropriate estimators, and their properties are now more complicated, and would require special investigation.

Some detailed proposals are made by Thompson (1992, pp 41–44) and Cochran (1977, pp 140–146). See also Hedayat and Sinha (1991, especially Chapter 12) and Levy and Lemeshow (1991, Section 3.5).

So far, in our example, we have assumed that we have stratified by one factor (here region) and have taken random samples from each stratum (region) over the other factor (here variety of apple).

But this is hardly realistic or economical.

If we stratify both by region and by variety of apple (an example of **multi-factor stratification**—see below) we could, of course, use the properties of simple random sampling for either interest, i.e. for regional means or for apple-variety means.

Alternatively, we might have stratified by region, chosen suppliers at random and *recorded simultaneously* in each case the prices $X, Y, \ldots$ of *each* of the varieties of apple on each sampling unit. We now have **multiple variables** of interest: our characteristic variable is now **multivariate**, and yet further difficulties arise. One problem is that what constitutes an optimum (or even a reasonable) allocation of stratum sizes for one variety of apple may not do so for another.

The first stage must be to determine and compare appropriate allocations for the different varieties separately. If the varieties do not differ widely in terms of variations in means, or in intra-regional variances, from one region to another, then a compromise can be adopted which is not far from the appropriate (optimum or, perhaps, proportional) allocation for each.

If they do differ widely (and with a wide selection of varieties this could be so) there may be no sensible compromise, and additional criteria must be introduced. These will usually be based on an assessment of the relative importance of the different variables. This might be a purely subjective procedure or occasionally a decision theory type of analysis might be conducted. The multivariable problem is widely encountered. Commonly we measure several outcomes on any selected unit in the population.

More formal methods of allocation with several variables of interest have been proposed. Some are discussed by Cochran (1977, Sections 5A.4–5A.6). One approach, due to Chatterjee (1967), considers the proportional increases in variance (4.28) from non-optimum allocation for the different variables of interest and seeks a compromise allocation which minimises the average value of these proportional increases. This can be reasonable if the proportional increases are not too widely different from each other. Otherwise, two early approaches due to Yates can be considered, based on defining a loss function for non-optimum allocation over the set of variables (Cochran, 1977, pp 121–123). These methods have been widely examined, particularly from computational algorithmic viewpoints.

Let us return to the notion of **multi-factor stratification**. In large scale surveys there will often be great appeal in stratifying the population with respect to several factors simultaneously. In a sociological enquiry it might appear desirable to divide the population into different sexes, different employment groups, different nationalities, different types of accommodation, and so on. Several practical interests support this: a desire to make the sample 'reflect' the population as a whole, or to facilitate the study of highly specialised subgroups (possibly different groups are of particular relevance for different variables being simultaneously recorded).

The *strata* become correspondingly specialised: female, highly paid, Welsh, lathe operators living in high-rise flats, and so on! For both the estimation of overall population characteristics, and of characteristics of subgroups, there will be some *disadvantages* in such *multi-way stratification*.

The individual factors of stratification (sex, income group, etc.) will have been chosen for practical interest. They may not necessarily correspond with desirable bases of stratification in the statistical sense of leading to improved efficiency of estimation of overall population characteristics. More seriously, simultaneous stratification by several factors soon leads to a vast number of strata—just 3 factors each at 4 levels yields $4^3 = 64$ strata. Except in a very large survey, individual stratum sample sizes are bound to be very small. The precision of estimators of stratum means (totals, etc.) will be correspondingly low. Administrative difficulties are likely to arise in the choice of random samples for the specialised combination subgroups. These subgroup (stratum) sizes and variances are unlikely to be known (or to be estimable) with any precision, so that the appropriate allocation of stratum sample sizes, or the estimation of the variance of estimators, will be difficult and unreliable. The widely used practical procedure of *quota sampling* (aiming to fill the 'quotas' for such complex combination subgroups) seeks to cope with these difficulties. See Section 4.7 below.

One particular problem is that there may be so many strata that it is not economical to ensure that they are all represented even with a single observation. Such a problem is discussed and illustrated by Cochran (1977, Chapter 5A), who once again provides a more detailed development of the various topics introduced throughout this sub-section.

4.6.4 *Post-hoc* stratification

Suppose plans have been drawn up to conduct a sample survey on a stratified population, and that stratum sizes and stratum variances are known. We can use the results derived in the earlier sections of this chapter to determine the variance of the stratified mean $\bar{y}_{st}$ for any allocation of stratum sample sizes—proportional allocation might be a typical choice. Alternatively we could calculate the optimum allocation that should be used to minimise $\text{Var}(\bar{y}_{st})$ or total cost.

However, the success of our efforts will rest on our ability to actually obtain the appropriate sample in the practical situation. We considered earlier some of the practical difficulties of drawing a truly random sample. In stratified sampling, we need such a sample *from each stratum*, and a major complication can arise in the respect that *we may not be able to determine in which stratum an observation belongs, until it has been drawn*.

This can happen, for example, where strata correspond to different personal details on people—such as their religious beliefs, income levels, nationality, educational achievement, and so on. For such factors, published national reports may provide a clear indication of stratum weights (sizes) and variances; but it can be most difficult to sample individuals from specific strata. One of the objectives of **quota sampling** (Section 4.7 below) is that it attempts to overcome such difficulties. Sometimes, however, we may have no alternative but to draw our sample and *stratify it subsequently*: that is, carry out a *post-hoc* **stratification**.

We assume there is no prospect of drawing a stratified random sample, since we cannot draw sr samples from specific strata. Instead we might take a sr sample of size n from the whole population, and subsequently assign individuals to the different strata. Although we do not obtain a stratified random sample, we should expect to encounter numbers in the different strata *roughly in proportion to the stratum sizes*, N_i. The resulting post-stratified sample should be somewhat similar to that which would have been obtained by stratified random sampling with proportional allocation, provided that numbers of individuals, n_i', falling in each stratum, i, are reasonably large. If we were now to estimate $\bar{Y}$ by the quantity analogous to $\bar{y}_{st}$ (rather than by $\bar{y}$), that is by

$$\tilde{\bar{y}} = \sum_{i=1}^{k} W_i \tilde{\bar{y}}_i,$$

where $\tilde{\bar{y}}_i$ is the mean of the observations that are found to lie in stratum i, we might expect to recover some of the potential advantages of $\bar{y}_{st}$ itself.

This proves to be so. As long as the S_i^2 do not differ widely, and the n_i' are large, $\tilde{\bar{y}}$ behaves similarly to $\bar{y}_{st}$ obtained from proportional allocation. Its variance is thus approximately $(1 - f) \sum W_i S_i^2/n$, which (as we have seen) can be considerably less than $\mathrm{Var}(\bar{y}) = (1 - f)S^2/n$. But how good is this approximation? It can be shown that the proportional allocation variance needs to be augmented by a factor to allow for the random (rather than fixed) stratum sample sizes to yield

$$\mathrm{Var}(\tilde{\bar{y}}) \approx (1 - f)\left\{\sum W_i S_i^2/n + \frac{N}{N-1}\sum(1 - W_i)S_i^2/n^2\right\}. \tag{4.30}$$

The form (4.30) is derived in Thompson (1992, pp 109–111).

Note that we need to know the stratum weights—otherwise we will need to resort to double sampling to estimate them (see Section 4.6.1). See also Cochran (1977, pp 134–135), Hedayat and Sinha (1991, p 276), Levy and Lemeshow (1991, pp 136–140) and Singh and Chaudhury (1986).

Another possible use of *post-hoc* stratification is to correct 'obvious lack of representativeness' in a sr sample, as illustrated in the following example.

Example 4.3

Suppose we draw a random sample of 10 individuals from the *Global Games* team data, and wish to estimate the population mean weight $\bar{Y}$. The particular sample drawn happens to contain 7 women, and 3 men; the sample mean is

$$\bar{y} = 62.40.$$

But we know that 60% of the population are men; our sample contains only 30% men and $\bar{y}$ is surely likely to under-estimate $\bar{Y}$ (which is 67.64). If instead of using $\bar{y}$, we work out the sample means for the men and women separately, we obtain, respectively,

$$\tilde{\bar{y}}_M = 80.33, \quad \tilde{\bar{y}}_F = 54.71.$$

The weighted estimator $\tilde{\bar{y}}$ for such *post-hoc* stratification yields

$$\tilde{\bar{y}} = 0.6 \times 80.33 + 0.4 \times 54.71$$
$$= 70.08$$

which is much closer to the true value (67.64).

Clearly, little can be claimed (other than vague intuitive appeal) for use of such a procedure in small samples, or without some prescription of what degree of 'lack of representativeness' will be needed to prompt the use of $\tilde{\bar{y}}$. But if the sample size is large enough to be confident that stratum sample sizes will also be large and if $\tilde{\bar{y}}$ is always used, whatever the constitution of the sample, we return to the earlier situation where $\tilde{\bar{y}}$ has essentially similar properties to $\bar{y}_{st}$ obtained by *proportional allocation*.

4.6.5 Choice of stratum boundaries for optimum allocation

Usually in a stratified population there will be many possible factors of stratification, e.g. sex, age-range, geographic region, ethnic group, etc. and the values of a variable of interest, e.g. *per capita* income will vary within and between the strata. The strata will typically be *pre-determined*; the values *fortuitously* varying within and between strata.

We know, however, that stratification is most profitable if values are relatively homogeneous within strata (S_i^2 ($i = 1, 2, ..., k$) small compared with S^2) and if stratum means vary widely.

So the question arises: if we have an opportunity to define the nature and number of strata at the outset rather than having to adopt pre-defined strata, how should they best be defined? The principle above implies that we should seek to define the strata in terms of the distribution of values of Y over the population—or of same variable highly correlated with Y. Essentially the choice should be one of partitioning the ordered population

$$Y_{(1)}, Y_{(2)}, ..., Y_{(N)}$$

(where $Y_{(i)} < Y_{(j)}$ if $i < j$) by means of $k - 1$ interior cuts, yielding k strata each of successively larger Y-values of varying sizes n_i ($i = 1, 2, ..., k$).

How this should be done for Neyman allocation was examined by Dalenius (1957) and reexamined by many others. Cochran (1977, Sections 5A.7–5A.9) presents and illustrates some of the results.

4.7 Quota sampling

Closely allied to stratified random sampling is the method of **Quota Sampling**, which is widely used in market research, opinion surveys, and a variety of nation-wide enquiries.

Its importance lies in the fact that it is the principal method of sampling employed by commercial data-collection organisations which service business needs for survey information, as well as being commonly used by individual agencies requiring regular re-appraisal of attitudes or activities in society. Political views, reactions to new or proposed government policy, patterns of trade and industry, consumer attitudes to products, and television audience sizes are all likely to be assessed through surveys based on quota sampling principles. The ubiquitous 'opinion poll' provides full scope for the method.

In essence, quota sampling is merely stratified random sampling with a complex multifactor stratification and with stratum sample sizes chosen by proportional allocation. The strata are chosen principally to ensure a 'representative picture' of the population with respect to the factors of stratification, and to yield estimates in specific subgroups, rather than primarily to enhance efficiency of estimation in a statistical sense. But efficiency enhancement can arise if the strata happen to have appropriate form (wide discrepancy of mean values, low internal scatter). The practical interest may often produce this effect—consider, for example, stratification by age, employment group, geographic region, etc., in different situations.

Where quota sampling differs from stratified random sampling is in the fact that the method of filling 'quotas' (numbers required in the different multi-factor stratum classes) can lead to the possibility that *stratum samples may not be random.* Typically, an element of subjective choice enters into the sampling practice because of the manner in which it is conducted.

Usually, strata are defined and the stratum sample sizes needed for proportional allocation are then calculated from (more or less) known overall stratum sizes in the population. The data are then collected by instructing interviewers or interrogaters to *fill the quotas* for the different strata, by street interviews, house to house enquiries, postal questionnaires, and so on.

The inevitable effect is that we cannot be sure that the selection of respondents is *at random* within the strata. To fill the quota by arbitrary selection would be time consuming. Successively more and more observations would be rejected as time goes on and as quotas fill up, and the practice is adopted of allowing the interviewer to 'use his or her judgment' to fill the quotas.

Thus we encounter in acute form many of the practical implementation problems in survey sampling—see Sections 1.2–1.5 and Chapter 6—how do we know if someone is a '40–50 year old professional male from the South East' and what we do if he will not answer our questions?

As time goes on more and more personal choice may be exercised in picking respondents—even style of dress can have an influence. To fill a quota of '40–50 year old, professional class men', the local doctor may well be deliberately omitted from a street interview if he is visiting the shops to replenish paint stocks, in the midst of painting the attic. This element of 'determined choice' means that *we cannot be confident in applying the results above in the quota sampling context.* In particular, non-response, which is inevitably ignored in quota sampling, can seriously bias results; consider (for example) television or radio audience assessments, where non-response might be highly correlated with viewing or listening behaviour.

This is not to say that quota sampling cannot produce very good results. *It can, and often does.* The difficulty is that we have no proper basis for measuring the properties of the sampling scheme since it is not truly probability based (see Section 1.6 and Section 6.3). Stephan and McCarthy (1958) give an early detailed appraisal of quota sampling methods and practice. See also Cochran (1977, pp 136–137) and, more recently, Curtis and Sparrow (1997), Lynn and Jowell (1996) and Sudman (1996).

4.8   **Ratio estimators in stratified populations**

In Chapter 3 we considered some of the ways in which the existence of an auxiliary variable X, 'correlated' with the variable Y of principal interest, can be exploited to provide better estimators of characteristics of the population of Y values than are obtained from a sr sample of Y values alone.

The *ratio estimator*, or *regression estimator*, provided this facility. One or the other was the more appropriate choice depending on the nature of the apparent relationship between the Y and X values (and depending on the extent of their association). If these estimators are to be recommended in simple random sampling from unstratified populations, there is reason to believe that the same may be true in stratified populations. To illustrate this we shall consider just one of the ways in which an associated auxiliary variable X can be employed in estimating characteristics of the population of Y values. The example we shall take is that of a *ratio estimator* of $\bar{Y}$ in a stratified population.

Suppose we draw a stratified random sample of values of (Y, X), with $n_1, \ldots, n_k$ observations in the different strata, the sample stratum means being, $\bar{y}_i, \bar{x}_i$ $(i = 1, \ldots, k)$. Suppose that in looking at the scatter diagrams of the data for each stratum there appears to be a fair degree of proportionality between the values of the two variables, shown in the form of a roughly linear relationship through the origin without substantial scatter. The slope need not appear to be identical in each stratum. Such an indication suggests that ratio estimators of stratum means or totals are likely to be profitable. We must assume, of course, that the stratum mean values $\bar{X}_i$, for the auxiliary variable, are known.

There are various ways in which we can combine such ratio estimators for the different strata to yield an estimator for the whole population. Two possibilities are to use the **separate ratio estimator**, or **the combined ratio estimator**.

Consider estimating $\bar{Y}$. The ratio estimator of $\bar{Y}_i$ is $(\bar{y}_i/\bar{x}_i)\,\bar{X}_i$, where $\bar{y}_i, \bar{x}_i$ are the stratum sample means.

Then the *separate ratio estimator* of $\bar{X}$ is

$$\bar{y}_{sst} = \sum_{i=1}^{k} W_i \frac{\bar{y}_i}{\bar{x}_i} \bar{X}_i,$$

i.e. the weighted average of the separate ratio estimators.

The *combined ratio estimator* reverses this process, forming *first* the stratified sample means and *then* correcting for the relationship between the two variables. It has the form

$$\bar{y}_{cst} = \frac{\bar{y}_{st}}{\bar{x}_{st}} \bar{X}.$$

The corresponding estimators of the total X_T will be

$$\sum_{i=1}^{k} \frac{\bar{y}_i}{\bar{x}_i} X_{iT}$$

(X_{iT} is the total of the X-values in the ith stratum), and

$$\frac{\bar{y}_{st}}{\bar{x}_{st}} X_T,$$

respectively.

How do these estimators compare? Both will tend to be biased unless the sample size is reasonably large. The bias will be less serious in $\bar{y}_{cst}$ than in $\bar{y}_{sst}$ since the sample size condition applies to the total sample rather than to each stratum sample. Approximate variances of $\bar{y}_{sst}$ and $\bar{y}_{cst}$ can be obtained by combining the results for the stratified mean and for the ratio estimator (typically (4.2) and (3.14)). It turns out that, unless the relationship between Y and X is the same in all strata, the *separate* estimator will be more efficient than the *combined* estimator. But this must be offset by the lower tendency to bias in $\bar{y}_{cst}$, and the fact that for this estimator we do *not* need to know the separate stratum means $\bar{X}_i$, only the overall mean $\bar{X}$.

The combined effect of stratification and the use of ratio estimators is somewhat unpredictable. It is appealing to think that potential gains from stratification should be further enhanced by using ratio estimators.

We can indeed *sometimes* do much better than either estimator separately; particularly when the auxiliary variable does *not* serve as the basis for stratification. But it is not as simple as this: the two effects are often tied up. Stratification can have the effect of reducing (even annihilating) the potential advantage of the relationship between Y and X, by weakening the relationship within the strata. The effects are by no means additive and the best we can say is that by combining the two techniques we should do as well as the better of the two on its own, for the problem in hand. Coupled with the extra effort and knowledge necessary if $\bar{y}_{sst}$ or $\bar{y}_{cst}$ are to be employed, this makes their use problematical.

Needless to say, all the separate practical and formal difficulties of stratified sampling, and ratio estimation, will enter into their use in combination (including problems of estimating variances, etc.). In terms of survey *design* the form of optimum allocation will tend to be somewhat different to that described in Section 4.4 when the ratio method is also used. See Cochran (1977, Section 6.14) and Levy and Lemeshow (1991, Section 7.7).

Analogous methods exist for using regression estimators in stratified populations, where these are more appropriate than ratio estimators.

4.9 Conclusions

This chapter has covered a large amount of material to do with sampling and estimation in stratified populations. To set the results in perspective, we conclude the chapter with a review of some of their implications. This is conveniently achieved by posing, and briefly answering, a few questions.

Why use stratified populations?

(i) In the hope of obtaining more efficient estimators than would be possible without stratification.

(ii) For administrative convenience; practical constraints of access or cost may compel different sampling techniques to be used for different sections of the population. The resulting data arise as random samples from the different sections; these sections constitute the strata in what is a stratified random sample.

(iii) Because we are interested in the sub-populations (strata) in their own right; or wish to 'represent' such sub-populations 'fairly'.

(iv) To reduce fortuitous bias in an unstratified sample, by *post-hoc* stratification.

Individually, or in combination, these factors support the use of stratified sampling, either by deliberate construction of appropriate strata (as in (i) and (iv)) or in an inevitable form determined by practical constraints and interests (as in (ii) and (iii)).

What are the advantages?

As implied by (i), (ii), (iii) and (iv), respectively, we would hope to obtain increased efficiency of estimation of population characteristics under appropriate circumstances, additional sampling convenience, greater ease of access to sub-populations of special interest or correction of unrepresentativity or 'bias'.

When does stratification lead to improved efficiency?

Population characteristics can be more efficiently estimated from a stratified sr sample than from an overall sr sample *if strata means differ widely, and within-strata variation is low*. The greater this effect the greater the efficiency of the corresponding estimators. With freedom of choice of strata, the aim should be to construct strata with these characteristics. If stratification is largely for administrative convenience the choice is limited and the efficiency improvement uncertain (although practical constraints do often produce a subdivision of the population appropriate to improvement of efficiency).

How should the population be stratified?

If unhampered by practical constraints, then clearly the aim should be to divide the population into non-overlapping groups of *Y*-values to maximise the separation, and internal homogeneity, of the strata. So the proper basis for stratification to achieve maximum efficiency is the set of *Y*-values itself. In practice, however, we will not have sufficient knowledge of the population to stratify it in this way. Instead we must employ some more tangible external criterion. If, as often happens, such a criterion corresponds reasonably well with separation of the *Y*-values into non-overlapping groups, little potential advantage from stratification will be lost.

For example, stratification by sex should prove a good criterion when estimating measures of physical stature; stratification by geographic region, or occupational category, likewise in estimating socio-economic factors. Although inevitably tempered by sampling ease and cost and the knowledge of stratum sizes and variances, we should strive to stratify the population in a way that is likely to produce the sort of stratification required from the theoretical standpoint.

How should the sample sizes be allocated to different strata?

Proportional allocation is particularly straightforward, and will often extract most of the potential advantages of stratified sampling. It is commonly used. If reliable information on stratum variances and sampling costs is available, then optimum allocation (or, for constant unit sampling costs, Neyman allocation) is to be recommended.

4.10 Exercises

4.1 A survey is to be conducted to estimate the total number of books borrowed from the 235 public libraries in a county authority during a particular week. It is possible to classify the libraries as *small*, *medium*, or *large* in size, on the basis of their stocks of books.

The numbers of books borrowed from libraries in the three groups are thought to be roughly in the proportions $1:2:3$. It is further anticipated that the within-group variances of numbers of books borrowed will be proportional to the square root of the corresponding means. There are 74 small, 133 medium, and 28 large libraries.

A total sample size of about 50 is required. If sampling costs in each group are the same, how should sample sizes in a stratified simple random sample be allocated to the three groups?

4.2 A stratified population has 5 strata. The stratum sizes, N_i, and means and variances, $\bar{Y}_i$ and S_i^2, of some variable Y are as follows.

Stratum	N_i	$\bar{Y}_i$	S_i^2
1	117	7.3	1.31
2	98	6.9	2.03
3	74	11.2	1.13
4	41	9.1	1.96
5	45	9.6	1.74

Calculate the overall population mean and variance, $\bar{Y}$ and S^2. For a stratified simple random sample of size about 80, determine the appropriate stratum sample sizes under proportional allocation, and Neyman allocation. Work out (for the same total size of sample) the efficiency of the sr sample mean $\bar{y}$ as an estimator of $\bar{Y}$, relative to the stratified sample means for the two methods of allocation.

4.3 A stratified population of total size N is made up of k strata of sizes N_1, $N_2, \ldots, N_k$. The ith stratum contains a proportion P_i of members possessing a particular characteristic ($i = 1, 2, \ldots, k$). If the fpc can be ignored, show that the variance of the stratified simple random sample estimator of the overall proportion, P, of population members possessing the particular characteristic is approximately

$$\frac{1}{n}\left\{\sum_{i=1}^{k} \frac{N_i}{N}\sqrt{P_i(1-P_i)}\right\}^2$$

when the stratum sample sizes are optimally allocated.

Suppose that a population of size 20 000 has 3 strata with weights 0.3, 0.6 and 0.1, and the stratum population proportions are 0.4, 0.7 and 0.2, respectively. A sample of size 120 is to be chosen. Determine the stratum sample sizes for equal allocation, proportional allocation, and optimum allocation.

Compare the efficiencies of the stratified simple random sample estimator of the population proportion for the three forms of allocation (the finite population correction can be ignored).

4.4 In a stratified population the sampling cost for obtaining a stratified simple random sample of size n, made up of n_i observations from the ith stratum $(i = 1, 2, ..., k)$, is

$$C = c_0 + \sum_{i=1}^{k} c_i \sqrt{n_i}.$$

The stratified sample mean, $\bar{y}_{st}$, is to be used to estimate the population mean $\bar{Y}$.

Determine the optimum allocation of the n_i (to minimise $\text{Var}(\bar{y}_{st})$ for fixed total cost C).

4.5 All the farms in a county are stratified by farm size and the mean number of hectares of wheat per farm in each stratum, with the following results.

Farm size (hectares)	No. of farms	Mean wheat (hectares)	Standard deviation
0–20	368	2.7	2.1
21–40	425	8.1	3.6
41–60	389	12.1	3.9
61–80	316	16.9	5.1
81–100	174	20.8	6.1
101–120	98	25.2	6.5
121–	138	31.8	9.1

For a sample of 100 farms, compute the sample sizes in each stratum under stratified simple random sampling with:

(a) proportional allocation;
(b) Neyman allocation.

Compare the precision of these methods (for estimating the mean number of hectares of wheat) with that of sr sampling.

5

Cluster sampling and multi-stage sampling

Sometimes a finite population may consist of a large number of groups of individuals, e.g. of households in a city. With this special form of stratification (many strata of rather small size) we refer to the strata as *clusters*. It can be cheap and convenient to draw a **cluster sample** as a sr sample of the clusters. If we observe all the members of the sampled clusters, this is known as **one-stage cluster sampling**. If we sub-sample from each of the selected clusters, we obtain a *two-stage* sample.

- We can estimate population characteristics using one-stage cluster sampling with equal sized clusters (5.1) determining the properties of the estimators from basic results for sr sampling.
- The results are readily modified (5.2) to cater for unequal sized clusters with estimators based on the **cluster sample ratio**, the **cluster sample total** and the **unweighted average of cluster sample means.**
- **Systematic sampling** can be usefully re-examined as a form of one-stage cluster sampling (5.4).
- Sub-sampling from the selected clusters (the primary units) to obtain secondary units (5.5) is known as **two-stage sampling**. **Multi-stage sampling** extends this principle into the area of **complex surveys**.
- We can obtain estimators of population characteristics from two-stage sampling with **equal sized clusters** (5.6) or **unequal sized clusters** (5.7) and determine their sampling means and variances.
- The closing commentary (5.8) stresses the administrative convenience of cluster sampling and multi-stage sampling; the exercises (5.9) cover various topics in the chapter.

We have already remarked on a major problem in survey sampling of matching the study population and target population. Sometimes there is no *list* to constitute our sampling frame and from which we can choose at random *individual members* of the population. Instead, the sampling frame often consists of a division of the target (or study) population into non-overlapping *groups of population members*. All population members are represented once only in these groups, which may be of different sizes.

There may be a convenient *list of the groups* in the population, which can be used to form the sampling frame.

It covers the population but its members (the *sampling units*) do not correspond to *individual* members of the population. Such loss of identification of individuals is offset by the great convenience of having a tangible list of sampling units in which to define a sample and, frequently, by practical advantages of cost or access in contacting chosen *sample* members (which are of course sampling units, or *groups of members* of the population).

For example, a list of addresses might be a convenient basis of access to individuals in households—but each address may correspond to several people. Or, in an enquiry into the performance of schoolchildren, choice of a sample of schools is likely to be easier and less expensive than choice of schoolchildren individually (irrespective of whether or not a complete list of schoolchildren exists).

Note how such dual considerations of the convenience of a list for specifying a required sample, and cost and access advantages, imply that we essentially sample from a *stratified population*. The strata are in fact the sampling units, which, in the sort of examples we have just described, are likely to be many in number, each containing relatively few population members. But methods of stratified sampling are unlikely to be appropriate. As described in Chapter 4 these involve *sampling from each stratum*. The administrative advantage in using a sampling frame with units representing many small strata lies in being able to *restrict the selection* of such strata. Thus instead of selecting some population members within *each* stratum, we will wish to select *some* strata but then to study *each* selected stratum *in full*.

This difference of emphasis is reflected in different terminology. The strata are now called **clusters**; the choice of a sample of such clusters to yield a sample of the population members is called **cluster sampling**. If *all the population members in each selected cluster* are used in the sample, the method is known as **one-stage cluster sampling**.

As a further development on this approach, we might draw a sample of clusters, but then sample only *some* of their members. This technique is called *sub-sampling*, or *two-stage cluster sampling*. Sometimes the members of the clusters themselves consist of groups of population members. The clusters are then called *primary units*, their constituent sub-groups *secondary units*, and so on. Selection of a sample by choosing from among the primary units, from secondary units within chosen primary units, and so on to further stages, is called **multi-stage cluster sampling**. For instance, in examining performance of schoolchildren in the educational survey referred to above, local education authorities may constitute the primary units, their schools the secondary units, classes within the schools the tertiary units, and children within the classes the members of the study population.

In the discussion of *stratified sampling* in Chapter 4, it was recognised that the manner in which a population is stratified may be conditioned by administrative factors. Nonetheless, the major interest in stratification was in its potential value *for producing more efficient estimators of population characteristics*. In contrast, *cluster sampling* is employed almost exclusively *for administrative convenience*; either to ease sample specification through the existence of a list of the clusters, or to improve access to the population, or to reduce sampling costs.

Sub-sampling

Cluster sampling methods are, and need to be, widely employed. Other methods, were they feasible, might well produce more efficient estimators but at much greater cost and administrative effort. The hope is that any loss in potential efficiency is outweighed by reduction in sampling costs, and greater sampling facility. Any objective comparison of cluster sampling with other methods needs to be made on this basis: it is unrealistic to exclude cost considerations in such a comparison, although the accurate specification of costs is not always easy, and it would be optimistic to claim that cluster sampling in practice is often supported by a justificatory cost analysis. Pragmatism is the major stimulus!

5.1 One-stage cluster sampling with equal sized clusters

As in our study of other sampling methods, we shall again restrict attention to just one or two aspects of cluster sampling. Detailed discussion will be confined, in the main, to one-stage and two-stage cluster sampling. Multi-stage schemes are, of course, often used and details of such complex designs are readily accessible elsewhere (e.g. in Levy and Lemeshow (1991), Lehtonen and Pahkinen (1995) and Skinner *et al.* (1989)). Note the difference between *multi-phase sampling* (e.g. see Section 4.6.1) and *multi-stage sampling* when in the former approach a sample is built up by drawing successively smaller samples to constitute the overall sample of required size. Multi-phase sampling does have a role to play in multi-stage sampling, for example in preliminary estimation of cluster variances or between-cluster correlations (see Section 5.6 below).

We start with one-stage cluster sampling. Suppose that the population consists of a set of *clusters* of individual population members. There are M clusters of sizes

$N_1, N_2, ..., N_M$ ($\sum_{i=1}^{M} N_i = N$). The members of the clusters are Y_{ij} ($i = 1, 2, ..., M$; $j = 1, 2, ..., N_i$), and the *cluster means* and *cluster variances* are $\bar{Y}_i$ and S_i^2 ($i = 1, 2, ..., M$), defined in the usual way. The population mean and variance are $\bar{Y}$ and S^2, respectively.

A sample of population members is obtained by taking a *simple random sample* of m clusters and including in the sample *all members of the chosen clusters*. The resulting sample, of size $n \geqslant m$, is a *one-stage cluster sample*. How are we to use it to estimate $\bar{Y}$?

The simplest case to study to begin with is where all clusters are of the same size. Suppose

$$N_1 = N_2 = \cdots = N_M = L, \quad \text{say.}$$

Then

$$N = ML.$$

The one-stage cluster sample, arising as a sr sample of m clusters, has size $n = mL$; the sampling fraction is

$$f = n/N = m/M.$$

Suppose the observations in the sample are y_{ij} ($i = 1, 2, ..., m; j = 1, 2, ..., L$).

As an estimator of $\bar{Y}$, the *cluster sample mean* might be considered. This is just

$$\bar{y}_{cl} = \frac{1}{mL} \sum_{i=1}^{m} \sum_{j=1}^{L} y_{ij}. \tag{5.1}$$

Since all clusters are of the same size, $\bar{y}_{cl}$ will be *unbiased* for $\bar{Y}$; that is

$$\boxed{E(\bar{y}_{cl}) = \bar{Y}}$$

Furthermore its variance is easily shown to be

$$\boxed{\text{Var}(\bar{y}_{cl}) = \frac{1}{m}(1 - f)\sum_{i=1}^{M} \frac{(\bar{Y}_i - \bar{Y})^2}{M - 1}.} \tag{5.2}$$

These results arise from the fact that $\bar{y}_{cl}$ can be expressed as

$$\frac{1}{m} \sum_{i=1}^{m} \bar{y}_i,$$

where the $\bar{y}_i$ are the cluster means for a sr sample of m of the M clusters. Regarding the complete set of cluster means $\{\bar{Y}_i, \bar{Y}_2, ..., \bar{Y}_M\}$ as the basic population from which we are sampling, this population has mean

$$\frac{1}{M} \sum_{i=1}^{M} \bar{Y}_i = \bar{Y}$$

and variance

$$\frac{1}{M-1}\sum_{i=1}^{M}(\bar{Y}_i - \bar{Y})^2.$$

The unbiasedness of $\bar{y}_{cl}$ as an estimator of $\bar{Y}$, and the form (5.2) for its variance, now follow directly from the earlier results (see (2.3)) for a sr sample mean.

Consider the alternative estimator of $\bar{Y}$ provided by the mean, $\bar{y}$, of a sr sample of $n(= mL)$ observations drawn without restriction from the total population (ignoring the cluster structure). This is also unbiased, and has variance $(1 - f)S^2/(mL)$. How does this compare in efficiency with $\bar{y}_{cl}$?

Note that

$$(ML-1)S^2 = \sum_{i=1}^{M}\sum_{j=1}^{L}(Y_{ij} - \bar{Y})^2$$

$$= \sum_{i=1}^{M}\sum_{j=1}^{L}(Y_{ij} - \bar{Y}_i)^2 + L\sum_{i=1}^{M}(\bar{Y}_i - \bar{Y})^2$$

$$= M(L-1)\bar{S}^2 + L\sum_{i=1}^{M}(\bar{Y}_i - \bar{Y})^2, \tag{5.3}$$

where

$$\bar{S}^2 = \frac{1}{M}\sum_{i=1}^{M}S_i^2$$

is the *average within-cluster variance*.

Thus

$$\text{Var}(\bar{y}) - \text{Var}(\bar{y}_{cl}) = \frac{1-f}{mL(M-1)}\{(M-1)S^2 - L\sum_{i=1}^{M}(\bar{Y}_i - \bar{Y})^2\}$$

$$= \frac{(1-f)M(L-1)}{mL(M-1)}(\bar{S}^2 - S^2), \tag{5.4}$$

by (5.3).

So the question of the relative efficiency of $\bar{y}$ and $\bar{y}_{cl}$ (in terms of a straight comparison of their variances, ignoring cost or convenience factors) has a simple resolution. *The cluster sample mean, $\bar{y}_{cl}$, will be better than the sr sample mean, $\bar{y}$, if the average within-cluster variance, $\bar{S}^2$, is larger than the overall population variance, S^2; and vice versa.*

It is interesting to observe that this is essentially the reverse of what was found in stratified sampling, where the method yielded greater efficiency if the within-strata variation was sufficiently *low*. Bearing in mind the basic difference in the two sampling techniques (cluster sampling and stratified sampling), this reversal is what would be expected intuitively!

There is another way of looking at these results. We can define a quantity called the *intra-cluster correlation coefficient*,

$$\rho = 2 \sum_{i=1}^{M} \sum_{j<k} (Y_{ij} - \bar{Y})(Y_{ik} - \bar{Y}) / [(L-1)(ML-1)S^2], \tag{5.5}$$

which provides an aggregate measure of the correlation between population members in the same cluster. This clearly has affinities with $\bar{S}^2$; the larger the value of ρ, the smaller, in general, we would expect $\bar{S}^2$ to be. In fact,

$$\rho = 1 - \left(\frac{ML}{ML-1} \right) \left(\frac{\bar{S}^2}{S^2} \right) \tag{5.6}$$

(see Exercise 5.2 at the end of this chapter), and we can express $\text{Var}(\bar{y}_{cl})$ as

$$\text{Var}(\bar{y}_{cl}) = \frac{(1-f)}{m} \frac{ML-1}{L^2(M-1)} S^2 [1 + (L-1)\rho]. \tag{5.7}$$

The condition, $\bar{S}^2 > S^2$, for $\bar{y}_{cl}$ to be more efficient than $\bar{y}$, now becomes

$$\rho < -\frac{1}{(ML-1)} \sim 0,$$

so that, as long as the population is large, the requirement for greater efficiency of $\bar{y}_{cl}$ is that the *intra-class correlation should be negative*. (Again, see Exercise 5.2 at the end of this chapter.)

It should be recognised that, since the prime stimulus for using a cluster sample is one of convenience, the greater efficiency of $\bar{y}_{cl}$ over $\bar{y}$, although feasible, is not likely to be widely encountered nor is it greatly important.

The population total Y_T can, of course, be estimated by $ML \, \bar{y}_{cl}$, and its variance is merely the appropriate multiple of (5.2).

In practice we will need to *estimate* $\text{Var}(\bar{y}_{cl})$, and an unbiased estimator is obtained by replacing $\sum_{i=1}^{M} (\bar{Y}_i - \bar{Y})^2/(M-1)$ in (5.2) by $\sum_{i=1}^{m} (\bar{y}_i - \bar{y}_{cl}^2)/(m-1)$. Under appropriate conditions we can again construct approximate confidence intervals for $\bar{Y}$, or Y_T, by assuming that $\bar{y}_{cl}$ is normally distributed.

Example 5.1

Consider, once again, the data on weights of athletes in the *Global Games* team.

There is no obvious practical basis for clustering in the team population but to illustrate effects we can construct 10 clusters each of size five as the consecutive groups of five athletes in the list shown in Table 1.2. Picking a cluster sample of size 2 we obtain an estimate of $\bar{Y}$ as $\bar{y}_{cl} = 71.80$ (compared with $\bar{Y} = 67.64$).

To examine the relative efficiency of the cluster sample mean, 1000 such cluster samples have been chosen and we see in Figure 5.1 how

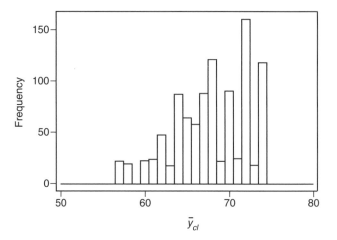

Fig. 5.1. Sampling distribution of $\bar{y}_{cl}$ for the *Global Games* team data

variable are the different estimates (note that there are only 45 possible values for $\bar{y}_{cl}$). The average value and variance of the values in the set of 1000 samples are 67.77 and 19.36 respectively. Thus we are encouraged to believe that the estimator is unbiased but it seems (not surprisingly in the terms of our discussion above) that it is not particularly efficient— the simple random sample mean (**Example 2.2**) has variance 13.62 whereas the ratio estimator (**Example 3.2**) has variance 7.62.

The same principles apply for estimating the *population proportion, P,* of Y-values satisfying some criterion. Suppose P_i ($i = 1, 2, ..., M$) are the corresponding cluster proportions. Our sample yields m of these; $p_1, p_2, ..., p_m$. The cluster sampling estimate of P is now

$$p_{cl} = \frac{1}{m}\sum_{i=1}^{m} p_i,$$

which is just a special case of $\bar{y}_{cl}$, where Y is assigned new values 1 or 0 depending on whether or not it satisfies the criterion of interest. The variance of p_{cl} is obtained from (5.2) by replacing the $\bar{Y}_i$, and $\bar{Y}$, by the P_i, and P, respectively.

We proceed to discuss what happens when the clusters have different sizes: a form of cluster sampling which is widely encountered.

5.2 One-stage cluster sampling with different sized clusters

Suppose as before that the population consists of M clusters, but that their sizes are $N_1, N_2, ..., N_M$ ($\sum_{i=1}^{M} N_i = N$) where not all the N_i have the same value. A cluster sample is again drawn as the basis for estimating some population characteristic, say $\bar{Y}$.

The cluster sample consists of *all* members of each of m clusters randomly selected from the M clusters in the population. Suppose that the sizes, means, and totals of the *chosen* clusters are $n_i, \bar{y}_i, y_{iT}$ ($i = 1, 2, ..., m$).

Certain complications now arise because the cluster sizes differ; various alternative estimators of $\bar{Y}$ might be considered, and their sampling behaviour is not always easy to determine precisely. We shall consider here just three possible methods of estimating $\bar{Y}$ (with obvious extensions to the estimation of Y_T or P); one further possibility is discussed later (Section 5.3).

It is useful to distinguish between the *primary units* (the clusters) and the *secondary units* (the population members within the clusters). The cluster sample is a sr sample of the primary units, and we can again carry over the results for sr sampling given in earlier chapters to study its behaviour.

We are essentially sampling a population of size M, where each member is represented by certain variables: for example, the cluster total Y_{iT}, the cluster mean $\bar{Y}_i$, or the cluster size N_i; for ($i = 1, 2, ..., M$). The cluster sample provides randomly chosen observed values $y_{iT}, \bar{y}_i, n_i$ ($i = 1, 2, ..., m$) of m of these variables.

The *overall* population mean $\bar{Y}$ (that is the mean value of Y over all secondary units) is

$$\bar{Y} = \sum_{i=1}^{M} \sum_{j=1}^{N_i} Y_{ij} \bigg/ \sum_{i=1}^{M} N_i = \sum_{i=1}^{M} Y_{iT} \bigg/ \sum_{i=1}^{M} N_i, \tag{5.8}$$

which is interpretable in the 'reduced' population of primary units as a *population ratio* of the total of the Y_{iT} values to the total of the N_i values. If we know both the number of clusters M, and the total overall population size $N = \sum_{i=1}^{M} N_i$, then writing

$$\bar{Y} = \left(\frac{M}{N}\right) \sum_{i=1}^{M} Y_{iT}/M, \tag{5.9}$$

$\bar{Y}$ is alternatively represented as just a known multiple of the primary population mean value of Y_{iT} (i.e. of the *mean cluster total* $\bar{Y}_T = (1/M)\sum_{i=1}^{M} Y_{iT}$).

These representations suggest two possible ways of estimating $\bar{Y}$ from the cluster sample.

(a) The cluster sample ratio

Use of the results of Section 3.1 on estimating ratios suggests estimating $\bar{Y}$ by

$$\bar{y}_{c(a)} = \sum_{i=1}^{m} y_{iT} \bigg/ \sum_{i=1}^{m} n_i, \tag{5.10}$$

which is the *ratio* of the sum of the cluster totals to the sum of the cluster sizes, in the chosen sample of clusters.

Thus (see Section 3.1) $\bar{y}_{c(a)}$ will have bias of order m^{-1} which will be unimportant only if the number of clusters in the sample is large. The variance of $\bar{y}_{c(a)}$ will be given by the approximation (see (3.5)).

$$\text{Var}(\bar{y}_{c(a)}) \doteq \frac{(M-m)M}{(M-1)m} \sum_{i=1}^{M} \left(\frac{N_i}{N}\right)^2 (\bar{Y}_i - \bar{Y})^2. \tag{5.11}$$

This variance depends on the variation between the cluster means: the smaller the variation, the smaller the variance. The effect is similar to what was found in the case of equal sized clusters; compare (5.11) and (5.2). $\text{Var}(\bar{y}_{c(a)})$ can be estimated from the sample by

$$\frac{(M-m)M}{m(m-1)} \sum_{i=1}^{m} \left(\frac{n_i}{N}\right)^2 (\bar{y}_i - \bar{y}_{c(a)})^2,$$

and if N is unknown, it may be replaced by the sample estimate Mn/m to yield

$$\frac{(M-m)m}{M(m-1)} \sum_{i=1}^{m} \left(\frac{n_i}{n}\right)^2 (\bar{y}_i - \bar{y}_{c(a)})^2.$$

To estimate the overall *total* Y_T we merely use $N\bar{y}_{c(a)}$; but here knowledge of the value of N is essential. The variance is just $N^2 \text{Var}(\bar{y}_{c(a)})$.

(b) The cluster sample total

If the total population size is known, (5.9) suggests that we might estimate $\bar{Y}$ from the usual sr sample estimate of a population mean. This implies using the estimator

$$\bar{y}_{c(b)} = \frac{M}{Nm} \sum_{i=1}^{m} y_{iT}. \tag{5.12}$$

Clearly,

$$E(\bar{y}_{c(b)}) = \frac{M}{N} \bar{Y}_T = \bar{Y},$$

so that $\bar{y}_{c(b)}$ has the advantage of being strictly *unbiased*.
Its variance is, from (2.3),

$$\text{Var}(\bar{y}_{c(b)}) = \frac{(M-m)M}{(M-1)mN^2} \sum_{i=1}^{M} (Y_{iT} - \bar{Y}_T)^2. \tag{5.13}$$

Since $\bar{Y}_T = (N/M)\bar{Y}$, (5.13) differs from (5.11) merely in the fact that the sums of squares of the cluster totals is calculated about the fixed quantity $M\bar{Y}/N$ rather than about the individual $N_i\bar{Y}$ which depend on the specific cluster sizes.

This implies that (5.13) will tend to be larger than (5.11), since cluster totals are most likely to be positively correlated with cluster sizes. This possible disadvantage needs to be set against the attraction of the unbiasedness of $\bar{y}_{c(b)}$. But the greater efficiency of $\bar{y}_{c(a)}$ is far from guaranteed: many factors enter into the comparison, including the relationship (if any) between the $\bar{Y}_i$ and N_i, and the variability of the cluster sizes. If the N_i do not vary greatly, then $\text{Var}(\bar{y}_{c(a)})$ need not be much different from $\text{Var}(\bar{y}_{c(b)})$ (indeed for equal sized clusters they are identical).

The estimation of $\text{Var}(\bar{y}_{c(b)})$ from the sample data, and corresponding results for estimating the population total, follow in the obvious way. Again there is the advantage that N need not be known for estimating Y_T.

Yet another estimator of $\bar{Y}$, which has as its principal attraction a very simple form which is easily calculated, is obtained as:

(c) The unweighted average of the chosen cluster means

That is, we use

$$\bar{y}_{c(c)} = \frac{1}{m} \sum_{i=1}^{m} \bar{y}_i. \tag{5.14}$$

This estimator is biased and inconsistent (in the finite population sense) unless all cluster sizes are the same. It provides a useful 'quick estimate' which will not be too seriously biased unless the cluster means and cluster sizes are highly correlated. Its variance is again obtained from (2.3), as

$$\text{Var}(\bar{y}_{c(c)}) = \frac{M - m}{mM(M - 1)} \sum_{i=1}^{M} (\bar{Y}_i - \bar{Y}_c)^2, \tag{5.15}$$

(where $\bar{Y}_c = (1/M) \sum_{i=1}^{M} \bar{Y}_i$) which can be estimated from the sample by

$$\frac{M - m}{mM(m - 1)} \sum_{i=1}^{m} (\bar{y}_i - \bar{y}_{c(c)})^2.$$

(If the bias is substantial, however, then the expected mean square error can of course be much larger than (5.15).)

The expected value of $\bar{y}_{c(c)}$ is $\bar{Y}_c$ so that its bias is

$$\bar{Y}_c - \bar{Y} = \frac{1}{M} \sum_{i=1}^{M} \bar{Y}_i - \frac{1}{N} \sum_{i=1}^{M} N_i \bar{Y}_i.$$

If the $\bar{Y}_i$, or the N_i, do not vary too much, this bias will not be serious, and the estimator can compare reasonably in efficiency with $\bar{y}_{c(a)}$ or $\bar{y}_{c(b)}$.

The corresponding estimator of Y_T is $N\bar{y}_{c(c)}$, with variance $N^2 \text{Var}(\bar{y}_{c(c)})$.

5.3 Cluster sampling with probability proportional to size

In Section 2.9 and elsewhere we have considered the implications of replacing sr sampling from the overall population with a sampling scheme in which the probability of choosing a sample member was not constant but was in some way related to its corresponding Y-value (where Y is the variable of interest).

Cluster sampling is another situation in which samples may usefully be drawn with different probabilities attached to the occurrence of different population members. This will be illustrated for one-stage sampling although the technique has (perhaps greater) advantages when applied to the choice of the primary units in multi-stage sampling.

Suppose we want to estimate the population total, Y_T, for a population which consists of M clusters of sizes $N_1, N_2, ..., N_M$ ($\Sigma_{i=1}^{M} N_i = N$) with cluster totals Y_{iT} ($i = 1, 2, ..., M$). For this purpose a single stage cluster sample, of $m < M$ clusters, is to be chosen. For simple random sampling the results have been presented in the previous Section. In many situations we will find that the larger clusters contribute the larger values Y_{iT} towards the total Y_T. It might seem sensible, therefore, to give greater attention to such clusters, perhaps by giving them larger probabilities of occurrence in the sample.

To keep the analysis fairly simple, we shall (as in Section 2.9) consider sampling *with replacement,* in such a way that the ith population cluster (of size N_i; $i = 1, 2, ..., M$) has probability p_i of being chosen. From the results of Section 2.9 we immediately conclude that the ideal value of p_i is Y_{iT}/Y_T.

The corresponding estimator of Y_T is then

$$\tilde{\tilde{y}}_T = \frac{N}{m} \sum_{i=1}^{m} \left[y_{iT} \frac{Y_T}{Ny_{iT}} \right] \equiv Y_T,$$

with *zero sampling variance.* But as before this is not feasible, *since if we knew Y_T we would not need to estimate it.* However, an associated alternative measure of cluster size *which may be known* is the number of population members, N_i, in the cluster. This suggests sampling the clusters with probabilities $p_i = N_i/N$, and the estimator becomes

$$\tilde{\tilde{y}}_T = \frac{N}{m} \sum_{i=1}^{m} \bar{y}_i, \tag{5.16}$$

which is just N times the average value of the means of the chosen clusters. Using an argument similar to that employed in Section 2.9, we see that this estimator is *unbiased,* and has variance

$$\boxed{\text{Var}(\tilde{\tilde{y}}_T) = \frac{N}{m} \sum_{i=1}^{M} N_i (\bar{Y}_i - \bar{Y})^2} \tag{5.17}$$

A corresponding estimator of $\bar{Y}$ takes the form

$$\tilde{\tilde{y}} = \tilde{\tilde{y}}_T / N = \frac{1}{m} \sum_{i=1}^{m} \bar{y}_i,$$

which is *unbiased* and has variance

$$\text{Var}(\tilde{\tilde{y}}) = \text{Var}(\tilde{\tilde{y}}_T)/N^2.$$

We can obtain unbiased estimators of $\text{Var}(\tilde{\tilde{y}}_T)$ and $\text{Var}(\tilde{\tilde{y}})$ from the sample, in the forms

$$s^2(\tilde{\tilde{y}}_T) = \frac{N^2}{m(m-1)} \sum_{i=1}^{m} (\bar{y}_i - \tilde{\tilde{y}})^2,$$

and

$$s^2(\tilde{\bar{y}}) = \frac{1}{m(m-1)} \sum_{i=1}^{m} (\bar{y}_i - \tilde{\bar{y}})^2.$$

The estimator $\tilde{\bar{y}}$ above is again described as being based on *probability proportional to size (pps) estimation*; it is a serious competitor to the three described in Section 5.2 above. Since the sampling has taken place with replacement, we must expect the procedure to be somewhat wasteful of effort: reflected in the value of Var($\tilde{\bar{y}}$). But even so, circumstances can arise (in particular when the $\bar{Y}_i$ and N_i are more or less uncorrelated) in which $\tilde{\bar{y}}$ has similar efficiency to $\bar{y}_{c(a)}$, and both are more efficient than $\bar{y}_{c(b)}$ or $\bar{y}_{c(c)}$. $\tilde{y}_T$ has the added advantages of unbiasedness and ease of calculation, although the sampling procedure is more difficult and can be more costly (in view of the emphasis on the larger clusters). A more detailed comparison of these estimators is given by Cochran (1977, Chapter 9A). See also Levy and Lemeshow (1991, Chapters 8–11) and Thompson (1992, Chapter 12).

Sometimes this approach may need to be modified. If the cluster sizes N_i are not known precisely, then it may be necessary to use an alternative measure of cluster 'size', perhaps an estimate of N_i or some other quantity likely to be positively correlated with the N_i. For example, suppose we wish to sample primary schools throughout the country by taking a one-stage cluster sample of local education authorities. If knowledge of the numbers of schools for each authority happened not to be readily available, certain other factors might well be: such as expenditure on school education in the regions covered by the authorities, or the total populations in these regions. Either of these will be highly correlated with the number of schools, and will constitute a reasonable basis for probability sampling. Such sampling is again referred to as **sampling with probability proportional to estimated size** (or **ppes sampling**).

So we now assume that each cluster has such a measure, Z_i, associated with it, and use this as the basis for *ppes sampling*, which consists of choosing m clusters, with replacement, where at each stage of drawing the sample, cluster i has probability $p_i = Z_i/Z_T$ of being chosen ($i = 1, 2, ..., M; Z_T = \sum_{i=1}^{M} Z_i$).

X_T and $\bar{X}$ are now estimated by

$$\frac{Z_T}{m} \sum_{i=1}^{m} y_{iT}/Z_i \quad \text{and} \quad \frac{z_T}{mN} \sum_{i=1}^{m} y_{iT}/Z_i,$$

respectively. These estimators are *unbiased*, and (see Section 2.9) have variances

$$\frac{1}{m} \left\{ \sum_{i=1}^{M} \frac{Z_T Y_{iT}^2}{Z_i} - Y_T^2 \right\} \quad \text{and} \quad \frac{1}{m} \left\{ \sum_{i=1}^{M} \frac{Z_T Y_{iT}^2}{N^2 Z_i} - \bar{Y}^2 \right\},$$

respectively.

In the light of what we saw of the effect of sampling with probability proportional to the cluster total, the most advantageous measure of size in *ppes sampling* will clearly be that which is most nearly proportional to the cluster total.

5.4 Systematic sampling revisited

Systematic sampling was introduced in Section 2.7. It involves taking observations at equally spaced intervals from a listing of the finite population. It is clearly a straight-forward and intuitively attractive sampling method but we also remarked on the possible dangers that might arise if there were systematic trends or groupings in the list. One difficulty was the fact that this approach does not necessarily correspond with sr sampling.

However, the sampling mechanism is in fact well-defined and it can be expressed as a form of stratified simple random sampling (see Section 2.7).

Interestingly, it can also be defined as a form of cluster sampling. *We are essentially taking a cluster sample of size m = 1!* We can see this as follows.

Consider a systematic sample chosen by taking one member at random from the first M on the list, and every Mth subsequent one. Suppose this yields a sample of size n. The population can be thought of as made up of M clusters:

$$Y_1, Y_{M+1}, Y_{2M+1}, \ldots$$
$$Y_2, Y_{M+2}, Y_{2M+2}, \ldots$$
$$\vdots$$
$$Y_M, Y_{2M}, Y_{3M}, \ldots,$$

of sizes differing by at most one. The systematic sample is just one of the rows drawn at random and is thus a cluster sample, consisting of one cluster chosen at random from the M clusters. Suppose $N = LM$, so that all clusters have the same size $L = (N/M) = n$; we can immediately apply the results of Section 5.1, with $m = 1$. If N/n is not an integer, so that the cluster sizes are not all exactly the same, minor modifications will be necessary in the terms described in Section 5.2 above, but the effects are qualitatively unaltered.

To estimate the population mean $\bar{Y}$ we take the *systematic sample mean*

$$\bar{y}_s = \frac{1}{n} \sum_{i=1}^{n} y_i,$$

where $y_1, y_2, \ldots, y_n$ are the observations in the single chosen cluster of the population. In the notation of Section 5.1, $\bar{y}_s$ is just $\bar{y}_{cl}$ based on $m(=1)$ clusters chosen at random from the $M = N/L$ 'systematic' clusters of size $L = n$ into which the population has been divided. So $m = 1$, $L = n$, $M = N/n$, $f = n/N$. We conclude that

$$\boxed{E(\bar{y}_s) = \bar{Y}}$$

so that $\bar{y}_s$ is *unbiased*, and by (5.2),

$$\boxed{\mathrm{Var}(\bar{y}_s) = \frac{1}{M} \sum_{i=1}^{M} (\bar{Y}_i - \bar{Y})^2 = \frac{n}{N} \sum_{i=1}^{N/n} (\bar{Y}_i - \bar{Y})^2,} \tag{5.18}$$

where $\overline{Y}_i$ is the mean of the *i*th cluster (or the *i*th of the *M potential* systematic samples). Equation (5.3) becomes

$$(N - 1)S^2 = M(n - 1)\overline{S}^2 + Mn \operatorname{Var}(\overline{y}_s),$$

so that if $\overline{y}$ is the mean of a sr sample of size *n*

$$\operatorname{Var}(\overline{y}) - \operatorname{Var}(\overline{y}_s) = \frac{n - 1}{n}(\overline{S}^2 - S^2).$$

The systematic sample mean will thus be more efficient than $\overline{y}$ if $\overline{S}^2 > S^2$; that is, in the event that the average variance within systematic samples is larger than the overall population variance. This can sometimes happen, and endorses the intuitive feeling that systematic sampling (as well as being easy to carry out) can be statistically advantageous: if systematic division of the population results in widely differing *Y*-values in each potential systematic sample. In practical terms this effect depends on the way in which the population has been listed (the listing may be highly structured in terms of *Y* values, or at the other extreme essentially random). Cochran (1977, Chapter 8) considers this matter in some detail.

The results can again be expressed in terms of the intra-cluster correlation ρ. We have, from (5.7), that

$$\operatorname{Var}(\overline{y}_s) = \left(1 - \frac{1}{N}\right)[1 + (n - 1)\rho]S^2/n,$$

and $\overline{y}_s$ has efficiency in excess of $\overline{y}$ if $\rho < -1/(N - 1) \sim 0$: that is (essentially) if ρ is negative.

If N/n is not an integer, some potential systematic samples will have one more member than others: in this respect the clusters in the population are not now all of the same size. The effect of such (small) differences in cluster sizes will be qualitatively unimportant unless the systematic sample size is very small.

Other specific topics in systematic sampling are covered by Hannan (1962) (review), Heilbron (1978), Valliant (1990) and Zinger (1980) (comparison of variance estimators, including stratified sampling); Hartley (1966) and Anthony (1983) (sampling with unequal probabilities); Bellhouse and Rao (1975) (effects of a trend in the listed values); Bromaghin and McDonald (1992) (systematic encounter sampling—see also Chapter 7); Dunn and Harrison (1993) (two-dimensional systematic sampling).

5.5 Multi-stage sampling

There are many ways in which the cluster sampling method may be modified or extended to cope with the specific demands of more complex situations.

One example arises where the selected clusters in the primary cluster sample are themselves *sampled*, rather than fully inspected. Thus if we choose a sr sample of *m* of the *M* clusters (*primary units*) which comprise the population, we may then take sr samples of sizes $n_1, n_2, \ldots, n_m$ of the *secondary units* in the chosen clusters. Thus we would now have $n_i < N_i$ ($i = 1, 2, \ldots, m$), in contrast to the sampling scheme of the previous sections where $n_i = N_i$. The process of drawing samples from the selected clusters is called **sub-sampling**: the resulting total sample of size $n = \sum_{i=1}^{m} n_i$ is called a **two-stage sample**.

If the secondary units are the individual members of the study population, there is no point in going further. If instead they consist of *groups* of population members, then we might either use all their constituent members, or proceed to further stages of sub-sampling. In this latter case we undertake **multi-stage sampling**, involving progressively higher levels of sub-sampling.

As described, the probability sampling mechanism at each stage is simple random sampling.

Multi-stage sampling can be illustrated for the school survey referred to in the opening pages of this chapter. In seeking a sample of school children, it could be particularly convenient to regard local education authorities as the primary units, schools under their control as the secondary units, and children in those schools as tertiary units. A *three-stage sample* (a sr sample of education authorities, a sr sample of schools under each of the selected authorities, a sr sample of children in each of the selected schools) has the advantages of being relatively easy to obtain, and of appearing to 'cover the population in a representative manner'. Additionally, without a complete listing of schoolchildren it would be a clumsy procedure to seek a direct sr sample from the whole population, whilst limited financial resources could imply that very few authorities (perhaps only one) would be chosen in a one-stage cluster sample, with the resulting risk of serious regional idiosyncrasies.

Thus the prime stimulus for multi-stage sampling is again administrative convenience, although scope exists for taking into account variance and cost considerations in the specification of what sizes of sample to take at the different stages.

Often sampling at later stages in a multi-stage design may involve stratification or use of concomitant variables (e.g. ratio estimators). A multi-stage survey of this type is known as a **complex survey**. Implementation, estimation of population characteristics and in particular variance estimation of estimators in complex surveys can be very complicated, involving detailed algebraic analysis and often the need to employ one of the currently available computer software packages. Modern survey samplings needs to enter this complex domain for full represention of the target population.

We shall not pursue the study of complex surveys in any detail. The texts by Skinner *et al.* (1989) and by Lehtonen and Pahkinen (1995) provide important introductions to the theme: relevant software packages include CLUSTERS and SUDAAN (see, for example, Lepkowski and Bowles (1996) and Scott and Wild, 2001).

For coverage of other general aspects of multi-stage sampling see Lee *et al.* (1989), Kish and Frankel (1974), Cochran (1977), Levy and Lemeshow (1991) and Thompson (1992).

Recent contributions to complex survey methods include problems of imputation (for missing data) which are discussed by Rao (1996), Chen and Shao (1999) ('hot deck imputation') and Chen *et al.* (2000) and in particular the use of sample re-use methods such as the bootstrap and jackknife (see Shao and Sitter, 1996, Shao and Chen, 1998 and Yung and Rao, 2000). Whitemore (1997) reviews multi-stage sampling with an emphasis on use of estimating equations; Pfefferman *et al.* (1998)

examine unequal selection probabilities and Shao *et al.* (1998) discuss variance estimation by balanced repeated replication. See also Sections 6.6.1 and 6.6.3 below.

5.6 Two-stage sampling with equal sized clusters

Suppose we have a population consisting of M clusters, each of similar size L, and we draw a *two-stage sample* by taking l members at random from each of a sr sample of m clusters. We assume that the cluster elements are the individual members of the population, and we wish to estimate the mean value $\bar{Y}$ of some measure Y defined on these members. The assumption of equal-sized primary units and equal-sized sub-samples *at this stage* simplifies the discussion.

The total sample size is $n = ml$; the sample members are y_{ij} ($i = 1, 2, \ldots, m; j = 1, 2, \ldots, l$); the within-cluster sample means are denoted $\bar{y}_i$ ($i = 1, 2, \ldots, m$). We denote by Y_{ij} ($i = 1, 2, \ldots, M; j = 1, 2, \ldots, L$), $\bar{Y}_i$ ($i = 1, 2, \ldots, M$), and $\bar{Y}$, respectively, population members, the cluster means, and the overall population mean.

The simplest estimator of $\bar{Y}$ is the analogue of the one-stage cluster sample mean $\bar{y}_{cl}$ (see (5.1)), namely

$$\bar{y}'_{cl} = \frac{1}{ml} \sum_{i=1}^{m} \sum_{j=1}^{l} y_{ij} = \frac{1}{m} \sum_{i=1}^{m} \bar{y}_i. \tag{5.19}$$

It is readily confirmed that $\bar{y}'_{cl}$ is *unbiased* for $\bar{Y}$, and that it has variance

$$\text{Var}(\bar{y}'_{cl}) = \frac{M - m}{mM} \sum_{i=1}^{M} \frac{(\bar{Y}_i - \bar{Y})^2}{M - 1} + \frac{L - l}{mLl} \sum_{i=1}^{M} \sum_{j=1}^{L} \frac{(Y_{ij} - \bar{Y}_i)^2}{M(L - 1)} \tag{5.20}$$

$$= \text{Var}(\bar{y}_{cl}) + \frac{L - l}{mLl} \bar{S}^2, \tag{5.21}$$

where $\bar{S}^2$ is again the *average within-cluster variance*.

As required, $\text{Var}(\bar{y}'_{cl})$ reduces to $\text{Var}(\bar{y}_{cl})$ if $l = L$, that is for one-stage cluster sampling (complete inspection of each selected cluster). When $l < L$, the variance is increased by the amount $(L - l) \bar{S}^2/mLl$, a contribution which arises from the further sampling variation due to subsampling the selected strata.

But care must be exercised in interpreting (5.21)! Whilst it literally declares that $\bar{y}'_{cl}$ has larger variance than $\bar{y}_{cl}$, this latter quantity requires full inspection for all m selected clusters so that the sample size is mL. But $\bar{y}'_{cl}$ is based on a sample of size ml which is typically *less* than mL. The fact that a smaller sample yields an estimator with larger variance is no surprise—in itself it does not reflect on the relative efficiency of estimation of $\bar{Y}$ in the one- and two-stage sampling situations.

The proofs of the unbiasedness of $\bar{y}'_{cl}$, and of the form (5.20) for its variance, are quite straightforward using conditional expectation arguments: taking within-cluster expectations conditional on the selected clusters followed by the marginal expectation with respect to the sr choice of clusters at the first stage. The details will not be presented.

The population total, Y_T, can be estimated by $N\bar{y}'_{cl}$. It is unbiased and has variance $N^2 \text{Var}(\bar{y}'_{cl})$.

In practice we will need to estimate $\mathrm{Var}(\bar{y}'_{cl})$. We can obtain an unbiased estimator as

$$s^2(\bar{y}'_{cl}) = \frac{M-m}{Mm} \sum_{i=1}^{m} \frac{(\bar{y}_i - \bar{y}'_{cl})^2}{m-1} + \frac{L-l}{MLl} \sum_{i=1}^{m} \sum_{k=1}^{l} \frac{(y_{ij} - \bar{y}_i)^2}{m(l-1)}. \tag{5.22}$$

At first sight, the second term in (5.22) contrasts strangely with the second term in (5.20): the divisor m in (5.20) has been replaced by M. The reason for this is that with incomplete inspection of the selected clusters,

$$\sum_{i=1}^{m} \frac{(\bar{y}_i - \bar{y}'_{cl})^2}{m-1}$$

is *not* an unbiased estimator of

$$\sum_{i=1}^{M} \frac{(\bar{Y}_i - \bar{Y})^2}{M-1},$$

and a compensation is needed which changes the divisor in the second term of (5.22) in the way described.

One case where this estimated variance (5.22) is particularly easy to obtain is where only a small proportion of the clusters is sampled. This is often the case in complex populations. Then the second term becomes negligible and we can estimate $\mathrm{Var}(\bar{y}'_{cl})$ simply from the values of the sample means of the selected clusters, as

$$\frac{M-m}{Mm} \sum_{i=1}^{m} \frac{(\bar{y}_i - \bar{y}'_{cl})^2}{m-1}.$$

Whilst the decision to take a two-stage sample is usually dictated by administrative considerations, some choice can often be exercised in the values of m and l. Within any limitations imposed by the total sample size, or total cost of sampling, different values of m and l will yield estimators with different variances, and the question of optimum choice arises. If substantial sampling costs occur only at the sub-sampling stage, and are the same in all clusters, then optimum choice of m and l for fixed total cost amounts to choosing m and l to minimise $\mathrm{Var}(\bar{y}'_{cl})$ for a prescribed total sample size ml. But such a cost structure is not often realistic.

When primary sampling and sub-sampling both involve costs, a more complicated cost model is needed. The simplest way in which such differential costs may be included is through a model which declares that there is a basic overhead cost d_0, that each selected primary unit costs an additional amount d_1, and each secondary unit an increment d_2, so that the total cost of sampling is

$$C = d_0 + d_1 m + d_2 ml. \tag{5.23}$$

Optimum choice of m *and* l now requires $\mathrm{Var}(\bar{y}'_{cl})$ to be minimised subject to the constraint (5.23) in which C is the total amount of money available for sampling. As in our study of stratified sampling, the dual problem of minimising the total cost of achieving a prescribed $\mathrm{Var}(\bar{y}'_{cl})$ does not require separate study.

The optimum value of l is the same in both situations, being approximately

$$\sqrt{d_1/d_2}\,\bar{S}(S_W^2 - \bar{S}^2/L)^{-1/2},$$

where $S_W^2 = \sum_{i=1}^{M}(\bar{Y}_i - \bar{Y})^2/(M - 1)$. The corresponding value of m may be obtained, as appropriate, from the prescribed value of C or of $\mathrm{Var}(\bar{y}_{cl}')$.

In practice, the slight departures from the optimum value of l which are likely to arise from the inevitable uncertainty of the values of d_1, d_2, $\bar{S}^2$, and S_W^2 are likely to lead to little loss of precision relative to the truly optimum choice.

When $S_W^2 - \bar{S}^2/L$ is sufficiently small, optimum choice of l will be L, so that *one-stage cluster sampling* is indicated as the best policy. This makes sense intuitively, since the variance of the cluster means, S_W^2, is now much the same as $\bar{S}^2/L$, with the implication that elements are assigned to clusters more or less at random. If so, complete inspection of a few clusters should be as efficient as partial inspection of many for the same total size of sample, $n = ml$. At the same time, complete inspection will, from (5.23), be less costly since $d_0 + d_2ml$ remains fixed whilst d_1m is minimised.

A much more detailed discussion of the properties of estimators and of the optimal choice of primary, and sub-sample, sizes in multi-stage sampling is given by Cochran (1977, Chapters 10 and 11), for this present situation and also for progressively more complicated situations (with unequal cluster sizes, stratified populations and more sophisticated probability sampling schemes including probability proportional to size sampling). See also Skinner *et al.* (1989), Thompson (1992, Chapter 13) and Lehtonen and Pahkinen (1995).

5.7 Two-stage sampling with unequal sized clusters

We shall consider here just one extension: two-stage sampling with unequal cluster sizes. For illustration of this situation, where we relax the unrealistic assumption of equal-sized primary units, we shall consider only the simplest case of a single estimator of the population total Y_T (or mean $\bar{Y}$) using sr sampling. Of course more complicated estimators are feasible, using the distinctions drawn in Section 5.2.

In practice we might also want to extend our coverage to estimation of proportions, to employment of some concomitant variable X observed along with Y, or to more complicated sampling schemes (e.g. with probability proportional to size). Estimating proportions poses no fundamentally new considerations, but the other complications take us beyond the scope of this present treatment. (Again see Cochran (1977), Lee *et al.* (1989), Kish and Frankel (1974), Skinner *et al.* (1989), Thompson (1992) and Lehtonen and Pahkinen (1995) for extensions of the coverage into some of these areas).

Thus, suppose we again have a population of M clusters, but now of possibly differing sizes $L_1, L_2, \ldots, L_M$. We draw a two-stage sample by choosing m clusters at random and then drawing sr sample of sizes l_i $(i = 1, 2, \ldots, m)$ from the chosen clusters.

Let the second-stage sample means and variances be $\bar{y}_i$ and s_i^2 $(i = 1, 2, \ldots, m)$; and the second-stage population means, totals and variances be $\bar{Y}_i$, Y_{iT} and S_i^2

($i = 1, 2, ..., M$). The overall sample consists of $n = \sum_1^m l_i$ observations which we denote y_{ij} ($i = 1, ..., m; j = 1, 2, ..., l_i$). The overall population has mean $\bar{Y}$, total Y_T, variance S^2 and size $N = \sum_1^M L_i$.

A commonly employed estimator of Y_T is

$$\hat{Y}_T = \frac{M}{m} \sum_{i=1}^{m} L_i \bar{y}_i \tag{5.24}$$

which is intuitively appealing—each $L_i \bar{y}_i$ estimates the corresponding cluster (first-stage unit) total y_{iT}; this is summed over the chosen clusters (first-stage units) and further enhanced by the factor M/m to provide an estimate of the overall total Y_T. It is easily confirmed that $\hat{Y}_T$ is unbiased.

An unbiased estimator of the variance of $\hat{Y}_T$ is given by

$$s^2(\hat{Y}_T) = \frac{M(M - m)}{m(m - 1)} \sum_{i=1}^{m} (L_i \bar{y}_i - \hat{Y}_T/M)^2$$

$$+ \frac{M}{m} \sum_{i=1}^{m} L_i(L_i - l_i)s_i^2/l_i. \tag{5.25}$$

If we denote by f_1 and f_{2i} the first-stage and second-stage *sampling fractions*, m/M and l_i/L_i, then the estimated variance (5.25) can be rewritten as

$$s^2(\hat{Y}_T) = \frac{N^2(1 - f_1)}{m\bar{L}^2} \sum_{i=1}^{m} (L_i \bar{y}_i - \hat{Y}_T/M)^2/(m - 1)$$

$$- \frac{N^2 f_1}{m^2 \bar{L}^2} \sum_{i=1}^{m} L_i^2(1 - f_{2i})s_i^2/l_i$$

where $\bar{L} = \sum_1^M L_i/M$.

Corresponding estimates for the population mean $\bar{Y}$ are readily obtained by using $\bar{Y} = \hat{Y}_T/N$ with estimated variance $s^2(\hat{Y}_T)/N^2$.

An alternative estimator of $\bar{Y}$ is

$$\hat{\bar{Y}} = \sum_{i=1}^{m} L_i \bar{y}_i \left/ \sum_{i=1}^{m} L_i \right. \tag{5.26}$$

which takes the form of a ratio estimate (see Chapter 3). It will be biased but this should not be serious if m (the number of sampled first-stage units) is large. A sample estimate of the MSE of $\hat{\bar{Y}}$ is given by

$$\text{MSE}(\hat{\bar{Y}}) = \frac{N^2(1 - f_1)}{m\bar{L}^2} \sum_{i=1}^{m} L_i^2(\bar{y}_i - \hat{\bar{Y}})^2/(m - 1)$$

$$+ \frac{N^2 f_1}{m^2 \bar{L}^2} \sum_{i=1}^{m} L_i^2(1 - f_{2i})s_i^2/l_i$$

which is basically similar in form to (5.25).

Yet another approach is to use *sampling with probability proportional to size* for the first-stage selection. A natural pps principle would be to sample with probability proportional to the size of the clusters (i.e. proportional to L_i) if this is feasible. See Cochran (1977, Chapter 11), Skinner *et al.* (1989) and Lehtonen and Pahkinen (1995) for more detail and for discussion of the relative efficiencies of the different methods.

Example 5.2

A videofilm hire company has shops in each of 5 regions: three regions have 12 shops, the others just 8. To estimate the total number Y_T of videofilms hired from the company in a particular week, the sales manager phones 12 shops chosen by picking 3 regions at random and then making a random choice of 3, 5 and 4 shops from the chosen regions.

The numbers of videofilms hired in the week were as follows:

first region:	260, 296, 182	(12 shops)
second region:	156, 261, 130, 302, 241	(8 shops)
third region:	196, 356, 268, 284	(12 shops)

Two estimates of Y_T are given by $\hat{Y}_T$ as in (5.24), and by $N\hat{\bar{Y}}$. We have

$$M = 5 \qquad L_i = 12, 12, 12, 8, 8 \qquad N = 52$$
$$m = 3 \qquad l_i = 2, 5, 4$$
$$\bar{y}_1 = 246 \qquad \bar{y}_2 = 218 \qquad \bar{y}_3 = 276$$
$$s_1^2 = 3396 \qquad s_2^2 = 5256 \qquad s_3^2 = 4309$$

So we obtain as estimates of Y_T

$$\hat{Y}_T = 13347 \qquad \text{and} \qquad N\hat{\bar{Y}} = 13013$$

with respective estimated standard error and root mean square error of 1640 and 810.

5.8 Comment

We have observed that the principal reasons for using cluster sampling techniques are administrative, rather than statistical, ones. The lack of a complete listing of population members, and differing problems of access to different groups of population members, may make it laborious, unfeasible, or costly to seek an overall sr sample from the whole population.

Very often the population has a hierarchical structure with successive levels containing smaller groups of population members, and access to the different levels is facilitated by the existence of lists of units at each level. The school survey was an illustration of such a structure. The most straightforward method of sampling, from the practical viewpoint, is likely to consist of drawing samples at each level: a sample of primary units, sub-samples of secondary units, and so on. In addition to such relative sampling ease, multi-stage sampling can also have the intuitive appeal of seeming to provide a more representative coverage of the highly structured population. Furthermore, the clusters may be of interest in their own right.

Statistical considerations enter with respect to the choice of what probability sampling scheme should be used, and what sizes of sample should be chosen, at the different stages. This choice will depend on the prevailing costs of sampling, but once we proceed beyond simple situations (such as described in Section 5.6), the appropriate analysis can be highly complex and even intangible, due to inadequate knowledge of the relevant cost and population variability factors although modern approximation and sample reuse methods, and supporting computer software, are making the implementation and analysis of complex surveys ever more accessible.

Even when a complete list of basic population members does exist, or can be compiled, it is often economically undesirable to take an overall sr sample from the whole population, and cluster sampling or multi-stage sampling is to be preferred. For example, the transport and administrative costs of drawing a sr sample of schools throughout Scotland (say) could be enormous. We could more economically select agents in 5 local education authorities, who can speedily and cheaply collect the required information on schools in their regions, and the savings in cost for such cluster sampling or multi-stage sampling will be large. It is unlikely that any loss of efficiency of estimation (for similar sizes of sample) could outweigh the economic considerations. Indeed the efficiency loss could be easily remedied by taking a *larger* cluster or multi-stage sample whilst still possibly retaining a substantial cost saving.

5.9 Exercises

5.1   A professional association publishes a list of its members. In a survey to estimate the average salary of members of the profession, this list is to be used as the basis for selecting a sample of about 5% of the 2640 members of the profession. Discuss difficulties and the possible advantages and disadvantages of employing simple random sampling or systematic sampling in the three cases:

  (a) where the list is in alphabetical order over all members,
  (b) where the list is in order of length of membership of the professional association,
  (c) where the list is grouped into different sections (e.g. employment sector, or nationality) and is in alphabetical order within each section.

5.2   Show that in a population consisting of M equal sized clusters, each of size, L, the intraclass correlation coefficient, ρ, defined by (5.6), can be expressed in terms of the population variance, S^2, and average within-cluster variance, $\bar{S}^2$, as

$$\rho = 1 - \left(\frac{ML}{ML - 1} \right)\left(\frac{\bar{S}^2}{S^2} \right).$$

Confirm the result (5.7) for the variance of the one-stage cluster sample mean, $\bar{y}_{cl}$. Discuss the implied restriction on the range of possible values of ρ, and show that

$$\frac{\mathrm{Var}(\bar{y}_{cl})}{\mathrm{Var}(\bar{y})} \doteq 1 + (L - 1)\rho.$$

5.3 A company, which provides its salesmen with cars for company business only, wishes to estimate the average number of miles covered by each car last year. The company operates from 12 branches, and the numbers of cars, N_i, and means and variances ($\bar{Y}_i$ and S_i^2) of miles driven last year (in thousands of miles) for each branch, are as follows.

Branch	N_i	$\bar{Y}_i$	S_i^2
1	6	24.32	5.07
2	2	27.06	5.53
3	11	27.60	6.24
4	7	28.01	6.59
5	8	27.56	6.21
6	14	29.07	6.12
7	6	32.03	5.97
8	2	28.41	6.01
9	2	28.91	5.74
10	5	25.55	6.78
11	12	28.58	5.87
12	6	27.27	5.38

Suppose that the average mileage is to be estimated by sampling a few branches at random, and using the figures for all the cars at the chosen branches. Work out the standard errors of the *unbiased* cluster sample estimator $\bar{y}_{c(b)}$ for a cluster sample of 4 branches, and of the sr sample mean for a sr sample of 27 cars (this being approximately the average number of cars which would be obtained in a cluster sample of 4 branches). Compare the efficiencies of the two methods.

What can be said of the use of $\bar{y}_{c(a)}$, or of $\bar{y}_{c(c)}$, in this situation?

Part 3
Practical aspects of carrying out a survey

6

Implementing a sample survey

The choice of a suitable probabilistic sampling design for a survey is only the first stage—it needs to be implemented. There can be many practical (largely non-statistical) decisions to make and difficulties to overcome.

- Variation of outcomes which are encountered is not always due to sampling (6.1); it may also reflect non-sampling *observational* or *non-observational errors*.
- Prior to drawing the survey sample it is usually necessary to take some preliminary observations or even to take a *pilot study* (6.2).
- There are many possible ways of collecting the survey data (6.3). These include *use of recorded data, direct observation, face-to-face interviews, postal (mail), web-based, e-mail or telephone enquiries*.
- The methods of obtaining the required information need special careful consideration especially if based on *questionnaires* covering *sensitive or personal matters* (6.4).
- The initial processing of the data requires choices to be made on *coding, scaling, editing and tabulation* (6.5).
- There are major problems to resolve (6.6) in *complex surveys* (where *replicated sampling* and *sampling reuse methods* can help). We need also to carry out appropriate procedures to handle *non-response* and *missing data*.

So far, we have considered in some detail the different probability sampling schemes, the forms and properties of appropriate estimators and how to achieve required accuracy or cost specifications. Such *statistical* considerations are crucial but it is also necessary to be able effectively *to carry out* any planned survey to meet required specifications. This leads to a range of (mainly) non-statistical matters to be determined and resolved. We noted the forms of some of these (e.g. how to actually obtain a random sample, the difficulties of non-response, etc.) in the introductory chapter (Section 1.2).

We shall expand on these non-statistical problems in this chapter, giving particular attention to:

- different sources of variation and error: natural, population access, response-related, etc.

- preparing the survey to face implementation difficulties
- data collection methods and associated problems: including different means of access
- obtaining the required information: ensuring confidence in responses, sensitive issues, ambiguities and misinterpretations of enquiries, questionnaire design, interview methods, non-response and its effects, etc.
- preparing to analyse the survey data
- dealing with specific problems of bias, misrepresentation and variance estimation: complex surveys, non-response, missing data and imputation methods, etc.

Such concerns are just a few of the practical matters which we must consider when moving from the 'drawing board' to the 'streetcorner'—indeed they also need to influence the final 'drawing board' (or design) decision on what type of survey is needed, with what sampling scheme, and of what size.

It is important now to examine this range of practical problems and to briefly consider how we might be able to deal with them or minimise their effects.

The material covered in this chapter could constitute a complete book in its own right. Indeed it has effectively done so on different occasions, and particular examples which usefully serve to extend our brief study over most topic areas are Hajek (1981), Hoinville, Jowell *et al.* (1978), Moser and Kalton (1971) and Sudman (1976). Selected references to wider discussion of particular topics are given in the appropriate sections. Some of these (such as questionnaire design, telephone and postal surveys, panel surveys, gross errors, effects of non-response or errors in response, complex surveys, etc.) have in fact—as we shall note—received extended textual coverage often as the subject of a whole book. We will give relevant references at the appropriate points in the text.

6.1 Sources of variation and error

This must be the starting point for our enquiries: what can go wrong, and why? Suppose that we have made *broad* decisions on how much to spend on a survey and on its statistical form. Before embarking on the survey we must obviously see if it is feasible to collect the data in the way specified by the design and what difficulties may be encountered in the process. This *pre-survey examination* (more fully described in the next section) will provide feed-back information from which we can if necessary mould the design into a feasible form.

The areas in which difficulties might be encountered can be readily categorised. A detailed coverage is given by Groves (1989) in his treatment of 'survey errors and survey costs'. See also Biemer *et al.* (1991), Lessler and Kalsbeek (1992) and Lyberg *et al.* (1997).

6.1.1 Sampling variation

Suppose we are again interested in some variable Y taking values $Y_1, Y_2, \ldots, Y_n$ over a finite population of size N, and in a characteristic of the population expressed as a summary measure of Y, e.g. the mean $\bar{Y}$. To know the whole population $Y_1, Y_2, \ldots,$ Y_N is to know everything about the population. We have seen that this is an unrealistic prospect and that we must seek instead to draw inferences (e.g. to estimate $\bar{Y}$) on the basis of a (probability) sample $y_1, y_2, \ldots, y_n$ of size $n < N$ where it is assumed that each observation y_i is a distinct member Y_j of the population of Y-values.

Obviously we hope that the sample provides a reasonable reflection of the population: in particular that it reflects the natural variation in the set of Y-values. This variation, sometimes called **sampling variation** (or **sampling error**) is inescapable: households *do* have different income levels, industrial companies do have different quarterly sales figures.

Why is this important? We have already seen that sampling variation (expressed perhaps in terms of the population variance, S^2) affects the accuracy with which we can, for a given sample size, estimate a characteristic such as $\bar{Y}$. Thus a sr sample of size n yields an estimator $\bar{y}$ with variance $(1 - f)S^2/n$; the smaller the sampling variation as reflected by S^2 the smaller, of course, is this variance. In practice we can estimate this sampling variance from our sample by $(1 - f)s^2/n$ for *post hoc* assessment of accuracy of estimation.

But we might want to specifically design the survey to achieve a prescribed degree of precision. This requires some advance knowledge of the value of S^2 and was discussed for simple random sampling in Sections 2.4 and 2.6 above. We also noted in Section 2.6 some of the ways in which an advance estimate of S^2 might be obtained to guide the choice of sample size, n, to achieve a prescribed degree of precision. These included use of *pilot studies*, consideration of the results of earlier 'similar' surveys, possible links between S^2 and $\bar{Y}$ implied by the physical situation and *double sampling*, (or *two-phase sampling*) in which our sr sample is made up of two sr samples, the first used to estimate S^2 and hence to determine an appropriate *overall* sample size to satisfy a given precision requirement (see Section 2.6). We shall consider *pilot studies* in more detail later (Section 6.2.1) and also return to the theme of *double* (and *replicate*) *sampling* in the contexts of complex surveys and of non-response errors (Section 6.6).

Such 'similarity' and 'model-structure' assumptions need to be cautiously applied, as often the two situations being contrasted are not as similar in practice as might be expected (e.g. short-term changes in social attitude can markedly affect personal expenditure patterns) or the expected links do not exist.

Thus it is important to be able to assess the extent of the natural sampling variation both for design of a survey and for interpretation of its results.

Example 6.1

An interesting example of a two-phase survey in the field of financial auditing is described by Barnett *et al.* (2001). In large-scale public or corporate audits it is not uncommon for the public or corporate body to carry out a large-scale random sample audit of their own transactions, from which an independent external auditor takes a random sub-sample of the internally chosen transactions and re-examines these transactions in detail. An extreme case might consist of an account (a population) of the order of one million transactions, with an internal audit of 100 000 transactions and an external audit of 1000 transactions.

It is assumed that the internal audit is prone to error—in the attribution of, and the assessed extent of, errors in the sampled transactions. It is further assumed that the external audit is totally reliable and precise.

Here we have a special form of two-phase, or double, sampling. It proves to be a complex matter to obtain optimum or effective estimators of the proportion, and the total monetary value, of errors in the account based on both the internal and the external audit data. These matters are addressed in detail by Barnett *et al.* (2001)—illustrating the importance of such a two-phase design but also the complexity of the analysis required to obtain the estimators in such an apparently modest extension of a sr sampling scheme.

For the classical sampling schemes of the earlier chapters, we have discussed in some detail how to estimate the effects of sampling variation. For more complicated schemes such as *multi-stage (complex) surveys* such assessment is more difficult and is beyond the scope of this book. Some aspects of the problem have been considered in Section 5.5 above and we will briefly examine a more general approach to variance estimation in complex surveys (based on replicated sampling and sample reuse methods) in Section 6.6.1 below.

Of course, natural variability in a population (sampling variation) should not be thought of as 'error'. It is an intrinsic and unavoidable feature of the population. But different types of error can occur when we *do not manage to achieve what the chosen sampling design requires*. There are essentially three types of such error:

- non-inclusion errors
- non-response errors
- observation errors.

Assael and Keon (1982) discuss the distinction between sampling errors and non-sampling errors.

6.1.2   Non-observational errors

We start with non-inclusion, and non-response, errors.

Non-inclusion errors, or **coverage errors**, arise when it happens that specific members of the target population cannot possibly arise in the sample. For example, a city street-corner interview survey is held on a Saturday afternoon when an important football match is taking place. This would be likely to greatly distort the accessible population and (depending on context) could seriously bias the survey results. Players and fans would not be encountered—but neither would individuals confined to hospital or prison! Then again a telephone survey cannot possibly cover population members who do not possess a 'phone. Basically, non-inclusion errors arise because of a serious mismatch of target population and sampling frame. It is tempting to say that such an obvious fault should not arise, but it is not always easy to anticipate all effects that might lead to the inaccessibility of some population members. Non-sampling errors in surveys is the theme of Lessler and Kalsbeek (1992). See also Groves (1989), Biemer *et al.* (1991) and Lyberg *et al.* (1997).

Obviously the bias arising from non-coverage errors will be greatest when there is a major distinction in the values, $\theta_{(c)}$ and $\theta_{(\bar{c})}$ respectively, of some characteristic of interest, θ, in the covered (accessible) part of the population and in the non-covered part. If $|\theta_{(c)} - \theta_{(\bar{c})}|$ is small, this implies that the non-coverage criterion is not closely correlated with the value of the characteristic and the effect on the bias of an estimator of θ will be less serious. For example, the absence of the Saturday-afternoon football fans may have much more effect on the variable 'age' than on the variable 'weight'.

Non-response errors are, as the name suggests, errors that arise from the fact that population members may be included in the surveyed sample *but do not yield a value of a variable of interest, Y*. Often we are simultaneously observing many variables; non-response may occur on *subsets* of the variables (e.g. age and income might not be declared) or on the *complete set* of variables (e.g. a questionnaire is not returned, in spite of reminders).

Non-response errors can arise for a *variety of reasons* related to the nature of the information sought (e.g. the facts or opinions sought), to features of the individuals in the population (e.g. whether persons, administrative units, industrial components), and to the method (or even time) of seeking to collect the information (e.g. whether by Internet, interview, questionnaire, telephone, panel or postal enquiry). Such factors are often interrelated: a member of a university group may be particularly reluctant to answer a personal question over the 'phone but might be more receptive to a face-to-face enquiry.

Often the *extent of* (*complete*) *non-response* is regarded as *a measure of the lack of success* of a survey but this is a dubious principle. Non-response may reflect unwise decisions on how, where, when and in what manner to try to collect the information and to some extent it can thus be controlled by sensible choice of operational procedures. But it is also recognised that different subjects of enquiry and different methods of collecting data inevitably produce different levels of non-response. Furthermore the non-response *rate* may not relate directly to the *extent of the error* that is engendered by the non-response.

The resulting loss of sample size will of course inflate the variance of estimators, but the degree of bias will depend on how typical (or atypical) is the non-responding group of the population as a whole. If the variable Y is strongly correlated with tendency for non-response (e.g. perhaps the more highly paid will be least inclined to reveal their incomes) we might expect to encounter serious bias as a result of non-response. We consider some statistical methods for handling non-response in Section 6.6.2 below.

Some commentators have attempted to classify types of non-response. For example, Cochran (1977, Chapter 13) distinguishes non-response due to intrinsic coverage errors, due to temporary absense from the sampling frame, due to being unable to understand or unwilling to respond to specific enquiries and due to adamant refusal to co-operate in the survey (the 'hard core' element).

The most difficult forms of non-response to cope with are personal non-responses due to *refusal to co-operate* (Cochran's 'hard core'); this is particularly influenced by the enquiry method (e.g. postal or personal interview) or even the very way in which an enquiry is framed. We will consider this in more detail below.

Failure to find selected sample members is another source of non-response and 'follow-up' or 'call-back' can help to remedy this. There is overlap here between non-response and non-coverage. Is the unlocated sample member really part of the sampling frame? This is partly a matter of definition. Consider the missing football fan in the city-centre, Saturday afternoon, street-corner survey!

Then there is the *located* sample member for which the required *information is just not available*, possibly because it is not relevant to that individual. Consider a traffic survey recording numbers of different types of vehicle at different sites over an observation period. No lorries are observed at one site, but lorries are banned on that observed section of road. Do we record 'no lorries' or 'non-response' or what? This is a common type of problem: distinguishing between 'not applicable', a zero-response and a non-response. Suppose a question 'How many hours television did you watch last week?' elicits a dash(—). Does this mean 'I didn't watch TV last week' or 'I'm not telling you!' or 'I do not have a television'? Such ambiguities are particularly prevalent on self-completed questionnaires. Failure to enter an answer is always a source of possible confusion for the statistician and careful survey design is essential to minimise its presence or influence.

6.1.3   Observational errors

Coverage errors and *non-response errors* are both examples of **non-observational errors**: errors due to non-observation. Equally serious are **observational errors**, where we obtain information from the chosen sample member but that information is faulty. This can happen in a variety of ways. A question may be misleading or wrongly expressed and lead to an incorrect response (we refer to this as an **interviewer error** or **question error**). An answer, although correct, may be wrongly recorded (a **recording error**), or subsequently wrongly coded (a **coding error**) or wrongly entered into a data-base (a **transmission error**). These errors are not the fault of the selected sample member.

In contrast, an individual may give an *incorrect reply* to even a well-posed question either deliberately to conceal information, or due to confusion not anticipated by the interviewer or survey analyst.

Such **response errors** occur if questions

- concern sensitive issues
- invite incrimination
- are over-detailed in structure
- are psychologically 'loaded'.

We shall examine these matters more in discussing methods of collecting the data, questionnaire design and dealing with sensitive issues in Sections 6.3, 6.4.1 and 6.4.2 below.

One further general class of errors needs to be considered. Termed **measurement error** (or perhaps **intrinsic error**) it arises when there is a specific value Y_i relevant to the individual i but it is physically difficult to observe without an additional superimposed error. Suppose Y is the pulse-rate of a patient. We measure it on the ith patient as 78. Does this mean $Y_i = 78$? Not really, since there will be error arising from the *inaccuracy of measurement*, and due to variation from moment to moment by natural (intrinsic) effects. Whilst in principle we might think of a unique Y_i, in practice we observe a random quantity: a random variable. The result we obtain might perhaps be usefully thought of as an observation y_i of a random variable Y'_i with mean Y_i, so that

$$y_i = Y_i + \varepsilon_i$$

where ε_i is the combined error effect of measurement and natural variation.

This is not an uncommon situation. Consider another example. In a traffic survey we want to study average daily flow rates over a year from different vehicle classes and types of road. Suppose Y_{ijk} is the flow rate for vehicle type i for the kth site of a type-j road. To observe Y_{ijk} (which is uniquely defined) we would have to sit at the roadside for the whole year. In practice, we will take counts over a much shorter time and seek to estimate Y_{ijk} from the limited-time data. Thus we will use a value

$$y_{ijk} = Y_{ijk} + \varepsilon_{ijk}$$

where ε_{ijk} reflects the estimation error.

We are now entering a branch of statistics which combines finite population methods with general random-variable-based statistical inference. This is a further illustration of the distinction between design-based and model-based approaches to survey sampling which we considered in Section 2.14. For further study of this topic, the treatment in Sections 13.8 *et seq.* of Cochran (1977) provides a useful introduction. See also Groves (1989), Lessler and Kalsbeek (1992), Lyberg *et al.* (1997).

To summarise the above description of the different sources of error we have the following categorisation.

General type	Forms	Specific versions
Sampling variation	sampling error	
Non-observational errors	{ non-inclusion error { non-response error	coverage error
Observational errors	{ response error	{ interview error, { question error { recording error, { coding error
	{ measurement error	{ inaccuracy of measurement, { intrinsic error

6.2 Pre-survey sampling

Many of the practical difficulties of conducting a designed survey can be assessed and evaluated by a modest amount of **pre-survey sampling**. Any data collected at this early stage can help to determine a number of critical factors, such as

- potential sources of measurement error
- likely non-response rates
- sensitive issues or sources of ambiguity
- interviewer inconsistencies
- difficulties of access to chosen sample members
- extent of variability (or some other characteristic) of some variable of principal (or secondary) interest.

No simple prescription can be given for the required extent of, or the method of collecting, such preliminary information other than to stress the need for appropriate randomisation and for avoiding obvious sources of unrepresentativity. How much pre-survey sampling is needed will depend on the range of problems which need preliminary investigation, as well as how much time and money is available for it.

Usually, pre-survey data are used to 'polish' the sampling design (e.g. as with the choice of sample size) and its method of operation (e.g. in the need for modification of questions, instruction of interviewers or allowance for various follow-up or 'reminder' stages). These data are not usually included as part of the main survey data to be analysed, although an exception might arise in the case of *double* (*two-phase*) *sampling* (or *replicated sampling*) see, for example, Section 6.6.2 below.

6.2.1 Preliminary work and pilot studies

The principal tool in pre-survey work is the *pilot study*. This can take many forms depending on its purpose.

At one extreme we may need to select just a few individuals on whom to try out different approaches to seeking information. This is particularly important on sensitive (personal) issues or where topics are complex and likely to be misinterpreted. *Extended, unstructured* and *in-depth interviews* with individuals or groups can provide useful preliminary information on which to design a questionnaire or 'approach procedure' for the main survey—perhaps even to indicate the potential relative merits of the different ways of seeking to obtain the data (see Section 6.3 below).

Provisionally chosen sets of questions and methods of sampling and of obtaining access will need 'polishing' by means of '*pre-tests*' of distinct aspects of the survey. For example if we are to conduct a survey on attitudes to local government spending based on face-to-face interviewing, we might need to test a section concerned with expenditure on ethnic issues. Will individuals understand the questions and interpret them properly; will interviewers be able to communicate adequately with the respondents; will different interviewers need different amounts of time?

Such matters of unstructured design work and pre-tests are discussed in more detail by Hoinville *et al.* (1978, Chapter 2), Moser and Kalton (1971, Chapter 2) and Levy and Lemeshow (1991, Chapter 15). See also Groves (1989).

At the opposite extreme, we may need to conduct what is effectively a minor version of the whole survey. The pilot study now takes the form of a **pilot survey**. Such pilot surveys, often based on random selection and possibly consisting of from 50 to (even) 500 sampled individuals, can be very important for the successful operation of a sample survey or opinion poll.

A pilot survey can be used for prior estimation of possible response levels and to elucidate the potential form of response errors. In turn, this enables more informed decisions to be made about the necessary sample size for the main survey and about the likely need for follow-up enquiries. It can also highlight the ways in which non-response or response error might be related to population features and could thus lead to biased and unrepresentative results. It enables useful preliminary comparisons to be made. For example, different covering letters could be compared in a postal survey or even different methods of data collection. A pilot survey can be used not only to examine non-response bias but to compare the cost efficiency of, say, a postal survey with a more structured 'drop-and-post-back' approach (see Section 6.3 below).

The pilot survey can also be used to obtain advance estimates of important summary characteristics of the population, such as the variance of a crucial measure (see, again, Section 2.6).

Other uses relate to more sophisticated sampling procedures, including *ratio* and *regression estimators* and *stratified sampling* (Chapters 3 and 4). In the former case, we noted that as well as sampling observations of a variable Y we must also observe a correlated (concomitant) variable X with particular advantages in terms of efficiency

Pilot surveys can help

of estimation of $\bar{Y}$, say, arising if we know the *population value*, $\bar{X}$. As an extreme example of a pilot survey we might be able to observe *all population values of X*, at modest cost, and hence evaluate $\bar{X}$ precisely.

Consider an industrial survey in which we are interested in an advance estimate of the mean level of sales $\bar{Y}$ in a particular sector for a financial accounting period just ended. Suppose full sales figures have to be returned in due course and are listed (as values of X) for the previous accounting period. It could be easy and cheap to use all these (as our 'pilot survey') to derive $\bar{X}$, and hence to enable a ratio estimate of $\bar{Y}$ to be obtained from a sample survey of Y-values in advance of the reporting stage.

With stratification, a pilot survey can provide vital information on which to base decisions on how to divide up the population (into its strata), on the difficulties of sampling the different strata and even on what sample sizes are needed in each stratum in the overall survey.

In all pre-survey activity, the notion of **fieldwork** is central: we must know how data collection methods will work out in practice. Will interviewers be able to achieve their objectives?; Do they need regular supervision?; Will it be feasible to obtain access to minority groups in the population?; and so on.

Pre-survey methods also have a vital role to play in estimating the likely differential and overall costs of carrying out the main survey.

Finally, the pre-survey stage is the one at which ethical factors are given proper consideration. These concern the relationships between the survey organisation and the respondents on the one hand and the clients on the other. Matters of confidentiality and intrusion are highly relevant as are the implications and requirements of formal legislation such as the *Data Protection Act* in the UK; see Cunliffe and Goldstein (1979).

It is important also to note how **disaggregation** may sometimes lead to a breach of confidentiality. If we disaggregate down to a level of fine detail even in a fairly large survey we may finish up with some cross-classifications containing only one or two sample members. Although not 'named' they can on occasions be readily identified from their presence in the particular cross-classification. This problem can arise when we are concerned with **small area estimation** (**subdomain estimation**) where we seek to estimate a population characteristic just for a small subgroup of the population. See Hedayat and Sinha (1991, Chapter 12) and Cochran (1977, Section 5A.13).

6.3   Methods of collecting the data

There are two distinct elements in the planning of a survey. One is the choice of the statistical sampling design, whether it is just sr sampling or one of the more complicated schemes that we have considered. The other is how we should carry out the designed survey: in the sense of collecting the data in accordance with the chosen sampling scheme. There are many possibilities, depending on the subject matter of the survey and the practical environment in which it will operate. We shall review the possibilities, covering specifically:

- recorded information (manual listings, computer databases, Internet records, etc.)
- observation

- face-to-face interviewing
- postal enquiries
- telephone interviews
- panel surveys
- snowballing
- e-mail and Internet.

6.3.1 Use of recorded or maintained information

The squirrel mentality is a feature of modern society! We collect everything, including information on all aspects of personal, scientific and societal action and interaction. This information is often recorded and may be available in different forms, at different levels of detail and with differing precautionary conditions, for the survey sampler to make use of. What larger resource can we imagine than all that is held on the World Wide Web!

Such data may have been compiled for administration and (local or national) governmental purposes, e.g. lists of names and addresses of members of the population (as in electoral rolls), levels of trading activity of companies of different types, 10-year National Census returns, vehicle licensing details. More specifically and more personally, information is held in medical records, police files, income tax returns, credit-worthiness evaluations, etc. Special groups keep membership details: from sociologists, through solicitors to soroptomists.

Finally, individuals keep their own personal records: as with the scientist's research work (or, perhaps, bank account details, letters to a close friend or collection of silver teaspoons!). Many Web entries reflect the urge to display information for others to see—a potential cornucopia for the survey scientist.

Such records might consist of just hand-written or printed hard-copy lists, or computer files or electronic listings on Internet sites.

In principle all such information could be useful—but will it be? Its use is limited by many factors. In particular,

(i) it may not really coincide with the target population of our survey
(ii) it may not be up-to-date
(iii) it may be too aggregated to be useful
(iv) it may be inaccessible for formal or legal reasons, or merely to protect the citizen (e.g. under the *Data Protection Act*), or may require computer passwords or registration for access to Internet sources
(v) groups or individuals may not co-operate in providing details that are sought, either at a personal level or for access to computerised records.

Web-based surveys are becoming widespread in certain countries. They take two basic forms: firstly a random sample is chosen of those linking to a specific site and then either:

- accessible information is recorded in the survey sample, or
- chosen contacts are invited to download and complete a survey questionnaire.

For current interests in the use of recorded or maintained information, only the first of these is relevant. It will provide only limited information in the terms of access data and what can be adduced by calling back to home-pages, etc., thus contributing only to basic descriptive surveys.

The second prospect allows a more structured survey to be carried out—subject to coverage or non-response problems—see Section 6.3.4 below.

Thus recorded information can be useful (e.g. electoral roll details) but often has limited value at the sampling-unit level beyond perhaps confirming (rather than providing) certain aspects of descriptive detail, e.g. sex, address. Occasionally its usefulness goes further, such as in an 'internal' survey where an organisation samples its own records to survey specific aspects of the characteristics of its members or internal structures.

Where it can be especially useful is in possibly providing a *list* or *sampling frame* on which to base the survey, or aggregate population details (sizes of sub-populations, orders of magnitude of means and variances of variables) to aid the *design* of the survey, rather than in providing the detailed information being sought in the *implementation* of the survey.

In recent years there has been a growing tendency for organisations to pass on (or sell) details of their recorded information to others for survey design purposes, e.g. a mail-order company may offer to sell a list of its customers details. This is a practice that is not universally liked by the recipients of unsolicited mail but is usually safeguarded by allowing individuals to opt out of such a provision. This opt-out further seriously limits the representativity and usefulness of lists or databases obtained from such sources.

6.3.2   Observation

It is natural to consider obtaining survey data merely by observing what is going on, without any need for communicative interaction. That is to say we look, rather than ask! This is a standard procedure for the scientist but may be relevant also to social or economic investigations. Although not widely used as a systematic principle for obtaining survey data, it has its role to play and we should briefly consider some advantages and disadvantages.

In some fields it is a common source of data. For environmental, biological and ecological studies it may be the only way, but it can be limiting. Often, only what is immediately available can be observed (Patil, 1991 and Barnett, 2002, talk of 'encountered data' in environmental studies; see Chapter 7 below) and what is 'available' may not be *representative* of the population we wish to study.

But observation-based data can on other occasions have the advantages of objectivity, accuracy and avoidance of response errors.

Even in a social enquiry about the way in which people might react to new situations, direct observation can be useful. For example in a marketing survey, a company might introduce a new product in trial areas and directly observe through points of sale how it is received (rather than asking by means of interviews how customers would like the new 'green' soap powder *if* it was introduced).

Again, scientific or technical data directly observed avoids the possible lack of understanding or knowledge on the part of respondents: e.g. as to whether their cars have servo-assisted brakes. Note that 'observation' might sometimes usefully be implemented (or augmented) by reference to published material (see above). The car manuals will say if the brakes are servo-assisted, but they will not, of course, reveal the extent to which different models feature in the target population.

Sometimes direct observation might be the only possible approach. Consider surveys on weed infestation in wheat crops, or abundance of species (or pollution levels) in rivers or ponds. We might need to go and look. Special approaches could be encountered in these examples. **Quadrat sampling** in biological surveys consists of randomly designating a sample area and counting what it contains, perhaps by throwing a standard one-metre square light wooden frame to define the sample region. **Inverse sampling** might be used for studying rare events: sampling at random until a determined number of events are obtained and noting the random number of trials needed to achieve this outcome. **Capture–recapture methods** involve taking a sample (e.g. of fish in a pond); marking them, releasing them and sampling again to observe how many marked individuals are obtained. See Chapters 7 and 8 for more details.

Direct observation can also greatly reduce the effects of non-response or of bias generated (often inadvertently) by the interviewer or responder.

Disadvantages are obvious. Direct observation can be time-consuming and highly costly; limited accessibility can also distort the nature of the very behaviour patterns we are hoping to observe. Also, it may not be possible. We can hardly expect to 'camp out' in households to watch and identify traits, attitudes or behaviour patterns of the household members—although this is not an unknown feature of anthropological or sociological surveys, particularly in closed communities of people or animals.

6.3.3   Face-to-face interviews

This is a common means of gathering survey data, particularly in opinion polls and attitude surveys. Armed (usually) with an appropriate questionnaire which needs to be well constructed (see Section 6.4.1 below), interviewers go out into the population and complete the questionnaire on behalf of individuals randomly selected according to the survey sampling scheme.

For successful and valid application this requires careful design of the questionnaire, fieldwork and evaluation by means of pre-survey work, engagement and training of interviewers (except for very small surveys), monitoring of their performance and assessment of the potential response of those interviewed to the prospect of face-to-face encounter. (Of course, not all of these factors are unique to the personal interview approach.)

It might be expected that this method will minimise misinterpretation and encourage high (if not full) response rate. A particular form is what is known as **quota sampling** (see also Section 4.7) where interviewers are instructed to constrain their 'random choice' of subjects by ensuring that specific numbers of particular categories of respondent are obtained. These constraints might be designed to achieve a particular *stratified sampling scheme* (as fully discussed in Chapter 4), e.g. a day's task might be to require the interviewer to obtain 20 male and 20 female subjects; in each case subdivided into 10 in each of social classes A and D, each of the 10 including 6 of age up to 30 years and 4 older than 30 years.

The potential advantage is that this can reduce (or even avoid) non-response. With street-corner interviewing, subjects will be approached until quotas are filled. But this has its dangers! How are the subjects to be chosen 'at random'? The interviewer has to decide if a potential respondent seems likely to be male and over-30 and in social class D! This implies a degree of subjectivity of choice, rather than random selection from the target population. At the end of a long hot day this can be particularly problematical. Then again *the avoidance of non-response can itself engender response bias.* How are we to be sure that those who agree to be interviewed do not have special attitudes that may be related to the very enquiries we are making?

Other difficulties in personal interviewing include perceived exposure or vulnerability, as well as general resistance to being interviewed, or a 'quiet-life' response ('yea-saying') on the part of the potential respondents. Thus, questions that may be acceptable in a private response situation may be resisted in a personal interview. Total refusal is also encountered 'on principle'. There is also the temptation to give responses which are thought to be required, or to be safe, rather than being true. The psychology of the responder–interviewer interaction is most complex.

An obvious feature of personal interviewing as a means of collecting survey data is its *inevitable expense*, through all stages of planning, training and implementation. It must be much more costly to send an interviewer out to ask someone questions, rather than looking up the information in records, sending out a questionnaire by post or making a 'phone call or e-mail enquiry. The justification must rest in factors such as ease of access, reduction of non-response and misinterpretation and (perhaps) the speed with which information is needed. Such factors are perceived to be especially important in **opinion polls**, and **market research**.

Needless to say, a personal-interview survey does not inevitably consist of casual street-corner encounters. A properly designed survey will often include a clear prior specification of the set of randomly chosen sample members to be interviewed with details of how to find them. This will involve determining a route and means of travel and a procedure to follow for those who are not found. Major cost implications enter into this aspect of the planning and implementation. In this mode, non-response cannot be reduced or eliminated but the inevitable **call-back** to those not found on the first 'trawl' can be expensive.

Another major problem is that of **interviewer errors**. The tendencies of respondents to give distorted answers (self-protection, self-esteem, 'easy-life') can, if we are not careful, be compounded by the interviewer. Ideally, the interviewer should not interact other than to provide neutral (non-subject-related) comments. This does not always happen. Biases can arise because of personal reaction of the respondent to the interviewer: prompting a desire to appear in a good light or, in contrast, a perverse intent to conceal or even mislead. There is also the risk of sheer misunderstanding by the interviewer of an answer that has been given. Sound and thorough training of interviewers is essential to minimise these difficulties.

Chapter 12 of Moser and Kalton (1971) and Chapter 5 of Hoinville, Jowell *et al.* (1978) provide interesting further details on questionnaires and interviewing as a basis of collecting survey data. See also Groves (1989), Hedayat and Sinha (1991, Chapter 11; Sensitivity Issues), Levy and Lemeshow (1991, Chapters 13 and 15), Fowler and Mangione (1990) and Peterson (2000).

6.3.4 Postal, web-based and e-mail surveys

An obvious alternative to the personal interview, and probably the most frequently used method, is to send out a questionnaire by post, by e-mail or through Internet (Web) contact, with a request that it is completed and posted back to the survey operator.

A hybrid approach of some interest is the '**drop-and-post back**' method, where questionnaires are *delivered personally* and left for completion and postal return. This has the potential advantage that the personal contact might engender interest and co-operation, enable an opportunity to explain difficulties and identify inaccessible potential respondents; with associated possible gains in response-rate and reliability. Of course, it is again an expensive approach (almost as expensive as personal interviewing) and only occasionally proves to be cost-effective.

With a **postal (mail)** or **e-mail survey**, all the usual design and planning procedures need to be followed (except usually for those involving interviewers). There may be an initial temptation for the recipient to throw the questionnaire 'in the bin' or not open the e-mail and efforts are needed to reduce this prospect. Thus, a well-constructed and persuasive cover statement can be useful (promising confidentiality), a stamped-addressed envelope is almost essential with a postal survey (it's hard for some to throw away an unused stamp) and tangible **incentives** may even be offered: 'return this form to claim your ball-point pen (or to be entered in our raffle)'. Dillman (1976, 2000) explores various aspects of the design and administration of (postal) mail and Internet surveys. See also Groves *et al.* (1988) and Mangione (1995).

In a Web-based survey there is a fundamental difference. If using mail or e-mail we can choose our target population (but see the constraints outlined below when using e-mail). For a Web-based survey the target population has to be accommodated within the set of those *who happen to make contact* with the Internet site running the survey. This even more severely restricts the form of the target population: all problems of representativity, coverage and response bias can be exacerbated and will need very careful controls to be in operation if the survey sample responses are to provide a sound basis for inference on the target population.

An important class of surveys does not in principle need incentives to encourage response. There are the many **official surveys**, where government (or other administrative or professional) agencies *expect* or even *formally require* their enquiries to be answered. Thus, companies may be required (if selected) to submit returns on their last-quarter trading to the relevant government department. But we should not imagine that this *guarantees* a full (or high) response rate. Government surveys seem in fact to encounter quite noticeable levels of non-response for various reasons (including perceived complexity of enquiry, or work load) and are regularly enough monitored and modified to seek to reduce this.

Postal surveys *can* lead to response rates as low as 50% or less. They inevitably require 'follow-ups' with **reminders** to non-respondents and often a further questionnaire and stamped-addressed envelope in each case. Often two reminders are sent. An old 'rule of thumb' claimed similar response rates at each stage: e.g.

Stage	*Overall response level*
First mailing	40%
First reminder	64% (40% + 40% of 60%)
Second reminder	78.4% (40% + 40% of 60% + 40% of 36%)

But it is of course dangerous to generalise! Response rates vary quite widely.

Other than reduction of costs, other advantages of postal and e-mail surveys are the elimination of interviewer errors and the opportunity to cover more detailed issues (perhaps requiring consultation between individuals or scrutiny of documents). But they are not necessarily a particularly speedy means of collecting data in view of the need to send reminders.

With the modern proliferation of home computers and Internet access it is tempting to view e-mail as a valuable and speedy means of sending survey enquiries. But there are some major problems. The coverage varies widely from country to country (the UK government has a policy of aiming for 100% Internet and e-mail access). Different age or social groups show varying degrees of interest, and involvement. Increasing levels of 'junk e-mail' are inducing a resistance to unsolicited e-mails (as with unsolicited postal mail) and this is heightened by the fear of computer virus infection from opening e-mail attachments.

Thus the attraction of cheap and speedy access by e-mail has to be set against the difficulties of covering the target population, of response bias due to the uneven mix of e-mail users and of resistence to the perceived intrusion of unsolicited e-mails. Such difficulties may became less serious as coverage and familiarity improve.

Nonetheless e-mail and Web-based surveys are to be found in all areas of public, social and commercial enquiry—a quick Web search of *mail surveys* and *e-mail surveys* threw up more than a million 'hits' in each case.

Much study and research is still needed to assess the relative advantages and disadvantages of postal (mail), e-mail and Web-based surveys. All encounter resistance, access problems and response bias possibly to an increasing extent as we go from postal (mail) to e-mail to Web-based modes.

Chapter 11 of Moser and Kalton (1971) and Chapter 7 of Hoinville, Jowell *et al.* (1978) give interesting discussions of postal surveys. More detailed treatments are by Erdos (1970) and Dillman (1976). See also Levy and Lemeshow (1991, Section 13.3). Couper *et al.* (1998) discuss 'computer assisted survey information collection' whilst Kennedy *et al.* (2000) give an interesting comparison of 'Web and mail surveys' concluding that 'currently, Web surveys are feasible [only] in constrained environments'. Some comparisons of postal (mail) and e-mail surveys are given by Tse (1998), Tse *et al.* (1995), Kittleson (1995) and Mavis and Brocato (1998). See also Dillman (2000).

6.3.5 Telephone surveys

With the rapid expansion of mobile phone usage it is ever more natural to think of try-ing to contact individuals (persons, organisations, institutions) in a survey 'over the 'phone'. This is not a recent idea. Even in the 1930s, in the United States, political opin-ion polls were being conducted this way but with some notable failures. One problem was that only about one third of the households then had a telephone. The effects are clear: low coverage of the population and high-response bias because of the inevitable non-representativity, in terms of social and economic characteristics, of these house-holds which had telephones (cf. e-mail or Web-based surveys at the present time).

The situation is of course very different now in coverage terms, with (perhaps many) more than 90% of households having a telephone in most developed countries: both at home and for personal use when away from home. It is not surprising, therefore, to find increasing use of the telephone to contact selected sample members in social surveys, opinion polls and market research.

The potential advantages of the **telephone survey** are obvious. It will be fast and cheap. Disadvantages include the non-coverage point outlined above. Even with 90% coverage, this will clearly not be random. Households without telephones are likely to reflect certain social and economic characteristics which in turn will be under-represented in a telephone survey sampling frame. Thus we can find notice-able *non-coverage bias* and this needs to be evaluated and assessed in each case.

Other difficulties echo these in the personal interview approach and relate to inter-personal effects between the telephone interviewer and respondent. This includes misinterpretation and misunderstanding of questions and answers, as well as emo-tional attitudes of protection, esteem or over-compliance (this latter often giving rise to what is seen to be the 'easy' answer, particularly in yes/no questions: phenomena known as '**yea-saying**' and '**nay-saying**'). As with all questionnaire-based enquiries it is essential that the questionnaire is designed to minimise such artificial responses (see Section 6.4 below). Equally important is the training of the telephone inter-viewers, following careful and detailed fieldwork and pre-survey investigations.

A feature which is probably more prevalent with this approach than with any other is the prospect of outright refusal to co-operate. People seem to particularly resent what they see as 'invasion of their space' by the telephone enquirer and special skills are needed on the part of the telephone interviewer to try to get past this first barrier. Sometimes this is not possible. Resistence to 'cold calling' has lead to the develop-ment of a registration system in some countries under which unsolicited calls are blocked. This exacerbates the coverage-bias and response-bias problems.

In spite of the problems, and because of the obvious time and cost advantage, increasing attention is still being given to telephone survey usage. One interesting development is Computer Assisted Telephone Interviewing (CATI). There is wide coverage of telephone surveys in the literature—see for example Dillman (1976, 2000), Groves *et al.* (1988, 2001) which survey the present situation in its various aspects. See also O'Neil (1979) on estimating non-response bias in telephone surveys and Bowan (1994) on reducing non-response bias when a postal survey is used as a 'backup' to a telephone survey.

Example 6.2

Mason and Traugott (2000) provide a most informative illustration of the implementation of a telephone-based interview survey. They provide a detailed practical example of the problems of non-sampling errors including non-response errors and the handling of missing data (see Section 6.6.3 below) reviewing and applying the range of current methods in a context of 'hard-to-reach' respondents and the need for multiple callbacks.

The survey discussed was a *random digit dialling* (RDD) telephone survey of adults in Oregon, US, where the RDD process generated relevant telephone numbers from working exchanges covering the target population. It was a multi-stage survey with geographic stratification and counties as primary sampling units with individuals chosen at random within selected (contacted) households. The survey was concerned with attitudes to a new political voting system. A total of 1483 interviews was completed: the overall response rate was 61.4%.

Many callbacks were needed both to fill in basic non-responses and to seek to reverse original refusals. There was also the need to cope with extensive missing data using appropriate 'imputation' methods. The broad conclusion of the authors to their extensive efforts to remedy non-response and refusals and to imputation of missing values is that 'persistence in working a sample to obtain a high response rate is unwarranted'. It may well be so in this case but such a conclusion is not universally true.

6.3.6   Other methods

For particular types of enquiry, there are other methods of data collection.

The keeping of **diaries** is a useful method when extensive information needs to be accurately compiled over a period of time. Typical examples are in *household expenditure surveys* and *television* (or *radio*) *audience measurement*. The disadvantages are the burden placed on the respondent and the expense of detailed monitoring and interaction.

Another time effect manifests itself in what are known as **panel surveys** (see Kasprzyk *et al.*, 1989); where the same set of sample members is maintained and observed at different points in time. The aim is to improve the accuracy with which we can examine the way in which a population changes over time: trade surveys or road-traffic surveys are typical examples for useful application.

In the **panel survey** the set of sample members is typically chosen in accord with a well-established, sample survey design. The ubiquitous '**focus group**' serving political or commercial interests is usually not so soundly based. It will tend to be a small group of individuals purposively chosen to reflect a wide range of interests and concerns which are judged to be important to the study in hand. It will not be a random sample and there is no statistical basis for extrapolating expressed views to a larger population. It is obviously hoped that the experience of those setting up the focus group would lead to its views being representative, and the focus group approach is vigorously defended (see, for example, Luntz, 1994).

There is a basic distinction of approach to be drawn between what are called **cross-sectional**, and **longitudinal**, studies. In the former (a cross-sectional survey) a single sample is considered at a fixed point in time (or over a fixed period of time) with the aim of estimating characteristics of the population *at that time*. In the latter (a longitudinal survey) similar enquiries are made at different times to examine the dynamic development of the members of the population. A *longitudinal survey* in which the same sample is used at each time is a *panel survey*, but there is no need for a longitudinal survey to have *total* coincidence of sample members at each stage. A diary survey is not *usually* longitudinal, e.g. a TV audience measurement survey is interested in the passage of time (a few weeks perhaps) as a basis for accumulating information rather than observing *changes* over time. A *series* of such *cross-sectional* surveys (**repeated surveys**) does of course show temporal development of population characteristics (e.g. average hours of sport watched per week). But compared with the case where the sample is fixed (and we then have a longitudinal panel survey) it does not enable us so readily and accurately to estimate population characteristics *of change over time*, e.g. saturation (or adulation) effects in the watching of soap operas!

6.4 Obtaining the required information

6.4.1 Questionnaire design

It will be clear from our discussions above that much of the success of a survey requiring objective and subjective responses from individuals and organisations rests on the skill with which a questionnaire has been constructed. So many factors can affect, *and distort*, the answers that are given. *Misunderstanding* is an obvious problem. Questions must be clear and unambiguous with response categories carefully chosen. This is easier said than done! Misunderstanding takes many forms. A question can just be too 'technical': the concepts are unfamiliar. It is easy to overestimate the vocabulary of the respondent. Consider the following question:

Q. *The government's efforts to reduce inflation have not been successful; it should persevere with these methods.*

Do you:	Strongly agree?	Agree?	Disagree?	Strongly disagree?

This could be a disaster area! The responder may not know what 'inflation' means— even 'persevere', which may be taken to mean 'discontinue'! But so many other problems arise with this question. It states a political viewpoint. Suppose the responder thinks that inflation *is* being reduced; this could modify the response (in anger, in defence?) compared with the answer that might be given to a neutral version of the question statement; e.g. *The government should continue with its efforts to reduce inflation.* Here we see the prospect for *attitudinal bias* in the response.

What about the prospect of a neutral response? The respondent is given no opportunity to 'sit on the fence' but must either agree or disagree. Do we want to force a directed response, or do we admit the valid prospect of no view either way?

In the latter case, we need an additional (centrally placed) neutral category, appropriately labelled. What about the labels? Should they go from 'strongly agree' to 'strongly disagree' or vice-versa? And what about possible effects of the '?' on each label, or even the different sizes of the response boxes?

An interesting forced and false response example was encountered over 60 years ago when responders were asked to give their views on which of a number of specific actions should be taken by the United States government in reaction to the international 'Metallic Metals Act'. About *three-quarters* of the respondents expressed firm views for action *in spite of the facts that no such Act had ever been passed or proposed* and *there was an explicit 'don't know' option*. Presumably this was a marked example of *self-protection*: individuals trying to hide their ignorance from the enquirer.

Thus there are major influences on the likely reaction of respondents to questions. They may modify their answers to *protect themselves* or *to comply* with what they feel is expected, or even (occasionally) *to deliberately confuse*. We can seek to anticipate these effects and reduce them by the design of the questionnaire. The types of questions (see below) and format for answers can be influential.

Personally sensitive questions present another problem area. If they are not central to the enquiry but required for classification of groups of responders we can sometimes use **proxy questions**. Rather than asking about social classes or income level, a few proxy questions on newspaper reading habits and location or type of residence can give a fairly clear estimate of these characteristics. This is not foolproof of course but may be adequate and enables the direct question to be avoided. If it cannot be, we have to use other methods. One is the **randomised response** approach. When first used it sought to estimate abortion rates by *asking at random whether the person had had an abortion or was born in a particular month*. The respondent knew that the enquirer did not know which question was being answered: minimising resistance to answering the question, avoiding legal reporting obligations but allowing for accurate estimation of the abortion rates (see next subsection: 6.4.2).

Other major factors that can irrelevantly affect, and possibly distort, responses are:

- the position of a question on the questionnaire relative to other questions,
- the very wording of the question,
- the interest of the respondent in the topic,
- need for memory in answering the question.

In Kalton *et al.* (1978) we see some interesting examples of the influence of question wording on question response. One simple example conveys a number of signals. In a survey of attitudes to local planning matters, respondents were asked 'Are you in favour of giving special priority to buses in the rush hour?'. 69% were in favour. In a repeated survey with different respondents those in favour dropped to 55% when the question was extended with the words 'or should cars have just as much priority as buses?'.

Questionnaire design is very important and is a major subject of research. Much has been written on it and the few remarks above can be usefully augmented by examining Chapter 3 of Hoinville, Jowell *et al.* (1978) or Chapter 10 of Groves (1989). See also Levy and Lemeshow (1991, Chapter 15), Lessler and Kalsbeek (1992) and the references at the end of Section 6.3.3. An extended treatment in relation to attitude surveys is given by Schuman and Presser (1981).

6.4.2   Sensitive issues: randomised response methods

It is common and inevitable that in survey-based social enquiries there will be questions that are, or are perceived to be, of a personal or otherwise sensitive nature. These may relate to medical or political issues. Such *sensitive issues* may lead the responder to refuse to give any response at all or (for example, in a questionnaire-based survey) to omit answers to some of the questions.

We have discussed in detail the reasons for *non-response* and ways of dealing with it (see Sections 6.1.2 and, later, Section 6.6.2). We will be further examining methods of coping with possible non-response in the specific field of environmental, ecological or medical enquiries in Chapter 7 below.

Partial non-response will also be considered in Section 6.6.3 where we discuss *imputation methods* for filling in the gaps left by *missing data*.

Such methods (other than those in Chapter 7) operate in a *post-hoc* mode in the sense that the procedures are applied *after* (at least some of) *the data have been collected.*

We will consider here a sampling method, the **randomised response method**, which seeks to deal intrinsically with the sensitivity issue by adopting a basic sampling scheme under which sensitivity is largely protected at the outset. It will apply to specific sensitive issues, which typically are not part of a larger enquiry. For example, suppose the aim is to obtain information about a medical or social condition (e.g. HIV or AIDS, political voting intentions, criminal records). Rather than asking a direct question on the sensitive issue, the survey is designed in a way where the answer which is given is not able to be unambiguously interpreted *for each individual* but where the overall set of responses allows inferences to be made on population characteristics for the issue of concern.

The method operates as follows. Suppose we want to estimate the proportion of a finite population who have a criminal record. The respondent is required to answer 'yes' or 'no' to one or other of two statements:

A   I do not have a criminal record
B   I do have a criminal record

but the respondent chooses which question to answer from the outcome of some randomising device, e.g. spinning a coin. The enquirer *does not know* which question has been answered and so the responder's privacy is protected and is further encouraged to respond accurately because of this anonymity. If by C we denote the condition in question **A**, the set of sample results can nonetheless be used to estimate the proportion

$P(C)$ of the population with that condition. Such a procedure is due originally to Warner (1965) and has been further refined in form and efficiency over the years; see, for example, Greenberg *et al.* (1969); Loynes (1976); Cochran (1977, pp 392–395) and the book by Chaudhuri and Mukerjee (1987).

How would we estimate $p = P(C)$? Suppose that the randomising device chooses question **A** with probability α (and **B** with probability $1 - \alpha$). Thus the probability of the answer 'yes' is ϕ, where

$$\phi = \alpha p + (1 - \alpha)(1 - p).$$

The value of α will be known; it is a feature of the survey design. Thus if r out of a sr random sample of n respondents from a finite population of size N reply 'yes', we can obtain an unbiased estimate of ϕ as $\tilde{\phi} = r/n$, with known variance (see Section 2.10). Hence from the form of ϕ above we obtain an unbiased estimate of p as

$$\tilde{p} = k[\tilde{\phi} - (1 - \alpha)]$$

where $k = (2\alpha - 1)^{-1}$ provided $\alpha \neq \frac{1}{2}$ (i.e. we could *not use* the unbiased coin as the randomising device!).

The variance of $\tilde{p}$ is readily determined as

$$\mathrm{Var}(\tilde{p}) = k^2 \mathrm{Var}(\tilde{\phi})$$
$$= (N - n)[p(1 - p) + \phi(1 - \phi)/(1 - 2\phi)^2]/nN$$

(see (2.18)) where the second term represents the increase in variance arising from the use of the randomising device.

A modification of this approach is known as **the principle of the unrelated (or irrelevant) question**. Here the second question is unrelated to the issue of interest, e.g.

B′   Is your birthday in June?

The respondent is even more comforted since it may not even be necessary to answer a question on the sensitive issue (see Horvitz *et al.*, 1967). If we know the percentage p_J of the population whose birthdays are in June then we can use the same principle as above to estimate p and determine its variance. (We can always find an unrelated question with a known percentage replying 'yes'; see Cochran, 1977, p 394.)

Again suppose that the randomising device has probability α of choosing question **A**. We have

$$\phi = \alpha p + (1 - \alpha) p_J$$

and can now estimate p by

$$p^* \doteq [\tilde{\phi} - (1 - \alpha) p_J]/\alpha$$

with variance

$$\mathrm{Var}(p^*) = k^2 \mathrm{Var}(\tilde{\phi}).$$

This will typically produce more efficient estimators of p than the straight randomised response approach using questions **A** and **B** (Dowling and Schactman, 1975).

Many modifications have been explored; e.g. use of two unrelated questions, estimation of the mean of a quantitative variable (see Greenberg *et al.*, 1971; Warner, 1971). The methods have been applied in many contexts such as voting intentions, abortions, drug use, criminal activity, etc.

6.5 Initial processing of the data

It is all very well to collect the survey data accurately in accordance with the chosen statistical (probabilistic) sampling scheme, but what should we do with it when it is to hand? The simple answer is that we need to construct the required estimates of population characteristics, together with estimates of their accuracy. With a modest-sized sample (a few hundred) and a few items of information, there is usually no difficulty. Most samples, however, arise from *complex surveys*, with large sample size and a multiplicity of inter-related measures: some qualitative or quantitative, some attitudinal or factual. Major statistical problems arise in estimating marginal and inter-relational characteristics and in estimating accuracies of estimates, e.g. in terms of their standard errors and correlations (see Section 6.6 below). Before this stage, however, there are more basic decisions to be made about the organisation and recording of the data for subsequent analysis with special attention nowadays to the requirements of computer systems and packages.

These matters include dealing specifically with:

- scaling
- editing
- coding
- tabulation
- missing data.

6.5.1 Coding and scaling

At the planning stage of a survey, decisions will have to be made on how to record the collected data. There are basic considerations, such as whether to maintain the data as completed questionnaires to be checked and analysed 'by hand' or whether to enter the results into a computer for detailed analysis. The latter is the more usual (and often the only feasible) approach. Data on questionnaires may be of two types: *structured responses*, to questions in which the respondent has to answer within a set of permitted categories ('place a tick in the relevant box, or boxes') or *unstructured responses*, to questions where a personal narrative response is invited ('What do you think of the management structure?').

For data processing (checking, editing, analysis and presentation) it is important to **code** the responses in a convenient manner. With factual structured questions, and particularly for computer storage and analysis, it is convenient to use numeric codes. Thus to an enquiry about the respondent's sex, we might code 'Male' and 'Female' as 0 and 1, respectively. Here the numbers have no significance; we could have used −2 and 17.1 but this would have been rather untidy. For a question on the terminal stage of education the permitted responses might be:

secondary, or high-school

college

university (first degree)

university (higher degree)

We could code these responses 0, 1, 2 and 3, but the choice is not now an arbitrary one. There is a natural ordering which would make 0, 3, 1, 2 rather perverse and even 3, 2, 1, 0 perhaps a little counter-intuitive.

If a question asks the respondent to state how many children there are in the family the coding is a natural one: 2 children is sensibly coded as 2 and the actual coded number is now meaningful, since it is the answer to the question. A choice might have to be made, of course, about grouping the responses. If we retain categories 0, 1, 2 and '3 or more', and code these as 0, 1, 2 and 3, say, then the coding is no longer fully numerically interpretable. Consider, for example, estimating the mean number of children by naively averaging the coded responses over all respondents.

With *structured attitudinal* questions, a (numerical) coding system will again be needed. It might be quite uninterpretable: no logical link between response and its coded number. Or it may need at least to reflect a natural order in that there is a valid sense in which progression through the coded numbers expresses a progression of attitude, e.g. strongly disagree, disagree, undecided, agree, strongly agree.

With *unstructured* questions, or those with large numbers of permitted distinct (and even multiple) responses, it may be necessary to determine coding schemes *post hoc*: deciding from the pattern of responses what might constitute a manageable set of response categories and a coding system for them. Obviously enough detail (i.e. a large enough number of categories) must be retained to reflect important areas of enquiry. It is unlikely, for example, that we could usefully contract 'number of children' into

0:   No children
1:   Some children

Such considerations should make it clear that there are two possible reasons for coding. The first is administrative convenience in processing the data, particularly by computer. The second is the possible interpretability of the numerically coded responses from the point of view of analysis. Even 'ordered' responses give some advantage in this respect. If the numbers are *fully interpretable*, e.g. the number of children, last year's salary, sales last quarter (even if grouped into a number of categories in terms, say, of mid-values either at the response or subsequent coding stage) then clearly they are particularly useful for statistical analysis purposes.

The ability to conduct meaningful statistical analysis is one of the stimuli to **scaling** of data. Its object is to transform a response (or a *combination of responses* to some related questions) to a numerical coding which is interpretable and will justify formal statistical analysis. It presupposes that there is a natural scale of measurement on which the qualitative responses (or sets of responses) lie, and seeks to determine what that scale is. This is a complex field of study and will not be pursued here (but see Chapter 14 of Moser and Kalton, 1971/1999, or Groves, 1989, Groves *et al.*, 1988, Levy and Lemeshow, 1991) other than to give a simple example of the problem. It might seem reasonable to code the responses:

> strongly disagree, disagree, undecided, agree, strongly agree

as 0, 1, 2, 3 and 4. At least there is an implied ordering preserved by this. But can we validly process such numerically coded responses in a statistical analysis? Does it make sense to say: 'the mean response is 2.7?' This depends on the underlying natural scale. To use the numbers 0, 1, 2, 3 and 4 makes highly specific assumptions about the relative values of different responses. Perhaps

> disagree, undecided, agree

are really rather close together, in which case

> 0, 2, 3, 4, 6

might be a much more reasonable scaling and coding system!

6.5.2   Editing and tabulation

When the data are available there are many processes of checking and editing that will need to be conducted. These include the inevitable, vital and time-consuming checking for *completeness of response*: is there an answer to each question on a questionnaire? If not, what should be done? *Missing data* is a serious problem and we shall return to it shortly. A fairly common difficulty is in interpreting whether non-responses imply no view, inapplicability, refusal to answer, or failure to ask the question. Apart from incomplete responses, the editing process must seek to high-light any *regular (or frequent) misinterpretations* evident in the pattern of responses, and seek to identify *specific inaccuracies*, perhaps evidenced by contradictory answers, e.g. affirmation of *both* 'no children' and 'oldest child is 10 years old or more'. But what about 'no children' and 'oldest child is less than 10 years old'?

Needless to say manual checking and editing can, and often does, go on at the same time as coding. But more routine editing is feasible and desirable when the data have been coded and entered into the computer. The computer can speedily conduct rapid and comprehensive searches for incompleteness and contradictions, including 'out of range' errors from coding, where a coded response is not included in the set for the particular question, and numerical responses which are statistically implausible in view of their extremeness (*outliers*). We return to the issue of faulty responses and missing observations in Section 6.6.3 below.

Before leaving the field of data preparation we should note that almost any detailed survey analysis will need to be preceded with a simple tabulation (ideally computer-generated) of the survey results. This takes the form of frequency tables of the responses to each question, and a judicious choice of marginal (one-way) counts and two-way cross-classifications.

The aim is informal: to receive broad indications of effects and relationships to guide the more formal statistical analyses. This is particularly pertinent for complex surveys (see Section 6.6.1).

6.6   Resolving problems particularly in complex surveys

6.6.1   Complex surveys: replicated sampling and sample reuse methods

We have referred already (in Section 5.5) to surveys in which a multiplicity of factors is being investigated, in a multiple cross-stratified population reflecting intricate population structure, and where accordingly very complicated sampling designs are needed. Of prime importance will be the study of relationships and measures of their strengths. This may need the use of regression and multivariate methods not usually employed in finite-population sampling studies. Skinner *et al.* (1989) and Lehtonen and Pahkinen (1995) provide details of such an approach. The basic problem with such **complex surveys** is in estimating variances of estimators: a topic specifically and extensively dealt with by Kalton (1977) and Wolter (1985) who included modern **sample re-use procedures** such as **bootstrap** and **jackknife** methods. A general sampling method which has advantages *inter alia* for estimation of variances is that of **replicated sampling** (or taking of **interpenetrating samples**). Here a sample is made up of a set of samples each obtained from the same (possible complicated) sampling scheme. Thus we have replicated samples each reflecting on (interpenetrating) the population, and intercomparisons can shed useful light on the sampling behaviour of (possibly complicated) estimators. To elaborate on these ideas it is interesting to consider an extension of the idea of double sampling, to that of **replicated** (or **multi-phase**; see also Sections 4.5 and 5.1) **sampling** in which, for example, a sample of size $n = cm$ is made up from c separate independent sub-samples each of size m and chosen according to the sampling design of the overall survey.

Such replicated sampling is one means of assessing the precision of estimators in arbitrary sampling designs, especially for estimating variances of estimators in *complex surveys*.

Suppose we choose to estimate a population characteristic θ by means of an estimator $\tilde{\theta}$ (a trivial example would be estimating $\bar{Y}$ by $\bar{y}$). Obviously, we need to be able to determine or estimate both the expected value and variance of $\tilde{\theta}$ to assess how good it is as an estimator of θ. In the simplest case of estimating $\bar{Y}$ by $\bar{y}$ from a sr sample then $E(\bar{y}) = \bar{Y}$ and $\text{Var}(\bar{y}) = (1 - f)S^2/n$ so that all we need is an estimate of the population variance S^2. But for more complex survey designs and more complicated estimators we may not know the theoretical form of $E(\tilde{\theta})$ or $\text{Var}(\tilde{\theta})$.

The replicated sample enables us to overcome this difficulty. Suppose each of the c sub-samples yields an estimate $\tilde{\theta}_j(j = 1, 2, ..., c)$ and we put $\bar{\tilde{\theta}} = \sum_1^c \tilde{\theta}_j/c$. Then we can clearly (trivially) estimate $E(\tilde{\theta})$ by $\bar{\tilde{\theta}}$ and an estimate of $\text{Var}(\tilde{\theta})$ is then directly provided by

$$s^2(\bar{\tilde{\theta}}) = \sum_1^c (\tilde{\theta}_j - \bar{\tilde{\theta}})^2/[c(c - 1)]. \tag{6.1}$$

The great advantage is that we can use this approach for *any form of estimator and any sampling design*. The choice of c is important and must balance the needs for

(i) many sub-samples to ensure high precision [via (6.1)], and for
(ii) sub-samples to be large enough to accommodate the sampling design (e.g. many *strata*: see Chapter 4).

Jackknife and **bootstrap methods**, as examples of what are known as **sample reuse** procedures, extend this approach. In *jackknife* methods we do not consider all c sub-samples of size m, but c *overlapping* samples of size $(c - 1)m = n - m$ made up of the whole sample dropping each of the sub-samples in turn. In the *bootstrap* approach, we might take repeated samples of size m from a larger sample of size n and explore empirically the sampling behaviour of particular estimators.

Kalton (1977) gives an interesting review of such ideas: see also Thompson (1992) and the comprehensive coverage of variance estimation in complex surveys by Wolter (1985), Lee *et al.* (1989) and by Lehtonen and Pahkinen (1995).

6.6.2   Non-response remedies

We have already discussed (in Section 6.1.2) some of the practical problems of non-response. Let us now consider how we might use statistical methods to remedy non-response bias and to estimate variances of corresponding estimators.

Suppose we seek a sr sample of size n, but obtain n_1 responses and $n - n_1$ non-responses. We might model the population as consisting (in principle) of R potential responders with characteristics $\bar{Y}_R$ and S_R^2 and $N - R$ non-responders with characteristics $\bar{Y}_{\bar{R}}$ and $S_{\bar{R}}^2$. If we could assume that the responses comprise a sr sample of size n_1 from the population of R potential responders, then if $\bar{y}$ is the sr sample mean,

$$E(\bar{y}) = \bar{Y}_R,$$

and the bias $(\bar{Y}_R - \bar{Y})$ can be expressed in the form $(N - R)(\bar{Y}_R - \bar{Y}_{\bar{R}})/N$. So the extent of the bias depends both on the non-response rate $(N - R)/N$ and on the difference between $\bar{Y}_R$ and $\bar{Y}_{\bar{R}}$. Furthermore

$$\text{Var}(\bar{y}) = (1 - n_1/R)S_R^2/n_1 \tag{6.2}$$

and the accuracy of estimation also depends both on the non-response rate and on the difference between $\bar{Y}_R$ and $\bar{Y}_{\bar{R}}$.

In the most favourable case, where non-response is unrelated to the variable Y, we have $\bar{Y}_R = \bar{Y}_{\bar{R}} = \bar{Y}$ and $S_R^2 = S^2$, and we have to compare (6.2) with $(1 - n/N)S^2/n$. But n/N is likely to be of similar order to n_1/R, so that (6.2) reflects an increase in the variance of $\bar{y}$ by a factor n/n_1. That is, non-response has effectively just reduced the sample size from n to n_1 and the loss of accuracy is expressible in terms of the (observed) non-response rate.

In practice, of course, the most favourable situation is unlikely to prevail and the comparison becomes more complicated. The notion of the population divided into two distinct groups (*strata*) of responders and non-responders is an example of a *stratified population* (Chapter 4).

Let us examine in more detail how the stratified model above can assist in resolving non-response bias. Double sampling can be particularly effective (Cochran, 1977, Chapter 13; Thompson, 1992, Chapter 14). Suppose again we have carried out a survey based on a simple random sample of size n and have obtained n_1 responses and $n_2 = n - n_1$ non-responses: regarding the population as stratified and made up of two strata of sizes R and $N - R$ of potential responders and non-responders, respectively. Suppose now however that we resample the original non-responders and obtain n_2' ($\leq n_2$) responses in the *second-phase sample.*

Effectively, we now have a stratified sample of size n' made up of n_1 observations from the initial potential responders stratum and of n_2' observations from the initial potential non-responders stratum.

Such data enable us to obtain unbiased estimators of population characteristics of interest and to determine and estimate their variances. For example, an unbiased estimator of the population mean is given by

$$\bar{y}_d = w_1 \bar{y}_1 + w_2 \bar{y}_2 = \frac{n_1 \bar{y}_1 + n_2' \bar{y}_2}{n_1 + n_2'} \tag{6.3}$$

where w_1 and w_2 are the stratum sample weights n_1/n' and n_2'/n', respectively.

This estimator is seen (via general results for *double sampling*; see Cochran, 1977, Chapter 12) to be unbiased. Using conditioning arguments its variance can be found as

$$\mathrm{Var}(\bar{y}_d) = \frac{(N - n')S^2}{Nn'} + \frac{w_2 S_2^2 (n_2 - n_2')}{n' n_2'} \tag{6.4}$$

where S^2 is the overall population variance and S_2^2 is the variance in the non-responder stratum. An unbiased estimator of $\mathrm{Var}(\bar{y}_d)$ can be readily obtained; see, for example, Thompson (1992, Chapter 14). See also Hedayat and Sinha (1991, Section 12.2) and Levy and Lemeshow (1991, Section 13.4).

The logic behind such an approach and the reason for using it is the belief that responders and non-responders may be intrinsically different in their response patterns, so that the simple rebalancing of sample size (described above) will not suffice. For example, in a postal survey on leisure activities people away from home for lengthy holidays, or for business trips, may well be different in their interests and patterns of leisure activity and expenditure than the 'run of the mill'. Missing out such respondents can seriously bias population estimates.

How then is the second stage sample to rectify the non-response pattern of the first stage? Why should we expect to be able to obtain responses from those who earlier failed to respond? (See Example 6.2.)

Often a different and more expensive sampling procedure may be needed—the initial postal enquiry may need to be replaced by repeated phone calls or by personal interview visits. Sometimes we cannot hope to 'patch in' a very high proportion of non-responses. On other occasions it is possible in principle to come as near to 100% recovery as we wish provided we spend enough time and money on the second-phase effort.

In this latter context, studies have been made of the implications of different cost models for the choice of an optimum proportion p of initial non-responders from whom we wish to ensure responses. An approach originally due to Hansen and Hurwitz (1946) and outlined in Cochran (1977, Section 13.6) shows, for a specified linear cost structure and a required final variance V for the estimator of the mean, how to effect an optimum choice of initial sample size n' and of $p = n_2'/n_2$ (the proportion of non-responses which become responses at the second-phase). Optimality is defined in terms of essentially minimising $C(V + S^2/N)$ where C is the total cost of the double sampling scheme.

6.6.3 Missing data and imputation

Much attention has been given in recent years to automatically (statistically) compensating for faulty responses and missing observations by seeking to 'predict' what might be reasonable 'substitute responses'. This is particularly prevalent in large-scale educational and social surveys. Many methods have been used including the sophisticated ideas of 'multiple imputation' (Rubin, 1987). There are differences of view on the routine use of such methods. Some argue that, if a large proportion of the data has to be dealt with, the process must be unreliable; whilst if only a small proportion requires attention it is probably not worth bothering!

There are different ways in which we might encounter **missing data**. In the usual multi-response surveys, we might have a full loss of an individual's response or a *partial response* where some enquiries have been answered and some not. The former is typical of what we have termed *non-response* (sometimes called *unit non-response*). Detailed discussion of its origins and forms is given in Section 6.1.2 above; Section 6.6.2 has explored possible statistical methods of correcting for non-response.

Partial non-response (or *item non-response*) is what is usually meant by the term 'missing data' which refers to the fact that certain fields are empty in the multivariate response vector of a selected individual. By definition some are not and it is this set of obtained information which might hold the clue to 'filling in the gaps' if this is our aim. An alternative is to omit any incomplete responses but this will contaminate the sampling design and the incomplete responses (as with non-response) may be correlated with other characteristics—thus leading to *incomplete-response bias*. Whether to omit incomplete responses will depend on what proportion of the sample they represent, on whether incomplete-response bias is anticipated and on whether the retained sample is large enough to meet estimation accuracy requirements.

If we decide to seek to 'fill in the gaps' in incomplete responses we have available a wide range of so-called **imputation methods**. Early developments are to be found in Ford (1983; specifically hot-deck procedures, see below). Kalton and Kasprzyk (1986) discuss the general treatment of missing data; see also Rubin (1987, multiple imputation methods) Levy and Lemeshow (1991, Section 13.5), Lessler and Kalsbeek (1992), Särndel *et al.* (1992) and Lehtonen and Pahkinen (1995).

Reasons for incomplete response are manifold: from failure to understand questions, reluctance to expose one-self to personal *sensitive issues* (possibly even with legal implications) to sheer fatigue ('this questionnaire is just too long'). Success or failure of imputation methods is bound to be tied up with the reasons for the incomplete responses. The range of prospects includes

- *deletion* of incomplete responses—with possible bias and imprecision
- *creating a new response field* 'unknown'—with possible interpretative difficulties
- *substituting the mean* of received responses (for a quantitative variable)—loss of person-specific information can be serious if broad-based marginal means are used.
- *hot deck imputation* where individuals are matched in terms of information that is present and missing data are filled in with subdomain means in the related (matched) groups—a *cold deck* approach uses subdomain means from *another* (related) survey
- *multiple imputation* (see, for example, Rubin, 1987) where each missing value is replaced by two or more imputed values to reflect the uncertainty of what value it is reasonable to use.

Mason and Traugott (2000) illustrate *inter alia* imputation of missing values in a telephone survey.

Part 4
Environmental sampling

7
Sampling rare and sensitive events

Often we are concerned with a population containing members that are *rare* and can be *difficult to identify*—or issues of a *sensitive or personal nature* on which it might be difficult to obtain reliable and correct information. We have also to cope, sometimes, with expensive observational and measurement costs. Many sampling methods exist for such problems.

- Specific concepts and methods (7.1) include *encounter sampling, screening, multiplicity sampling, disproportionate sampling* and *snowball sampling.*
- *Composite sampling* (7.2) can be used to examine economically and confidentially if rare members are present and to estimate the proportion of them in the population.
- In *ranked set sampling* (7.3) we can efficiently estimate population characteristics when sampling costs are high, by deliberately 'spreading out' the observations. In the process we can achieve efficiencies well in excess of those obtained from simple random sampling.

There are a many situations in which the aspects of a population which we wish to survey are *rare* and can be *difficult to identify*—they may also relate to issues of a *sensitive or personal nature* so that even when a relevant population member is identified it may be difficult to obtain reliable and correct information.

Examples are enquiries or studies about the presence in a population of minority groups (golden eagles, genetic variants or 'gaol birds'). The latter group members may, as well as being relatively rare, also be dis-inclined to reveal their criminal record. So to rarity we have to add the complications of overcoming resistance to answering enquiries even if members of the group are found. Another example of rare events arises in statistical auditing of financial accounts; we considered one aspect of this in Example 6.1.

We have already considered different aspects of these problems, as in the discussion of questionnaire design (Section 6.3.1) and non-response (Sections 6.2 and 6.4.2) where the aims were to overcome reluctance to respond accurately and fully, especially on sensitive issues. Specifically on this latter topic, we discussed in Section 6.4.2 the randomised response method for collecting data on such sensitive matters.

We will encounter the need to examine rare, or sensitive, matters in a variety of contexts but they will be particularly prevalent in biological, ecological, environmental, epidemiological (medical) and societal contexts. In this chapter we will consider a range of further methods for revealing information on rare events and sensitive issues.

Another complication we will consider in this chapter is the problem of collecting sufficient data when observation and measurement of individuals may be particularly *time-consuming and expensive.* An extreme example was referred to in Section 1.5 where in a survey of the assets of a water company it was noted that individual observations could cost £250 000 or more! Obviously we would need in such cases to keep sample sizes as small as possible; equivalently to use the most efficient sampling scheme possible. In the many areas of environmental risk such as radiation (soil contamination, disease clusters, air-borne hazard) or pollution (water contamination, nitrate leaching, root disease of crops) we commonly find that individual measurements can involve substantial scientific processing of materials and correspondingly high attendant costs. So again it becomes particularly important that we should seek to draw statistical inferences as expeditiously as possible with regard to containing the sampling costs. More detailed coverage of many of the topics in this chapter and in the next (in the context of 'sampling natural phenomena') is to be found in Thompson (1992) and in Barnett (2003).

In the fields described above, particularly in relation to environmental and some medical or social enquiries, the opportunities to use the sample survey methods described in the earlier chapters will sometimes arise. Often, however, *this will not be so* since there will be no opportunity to use simple random sampling as the basis of our data collection process. Instead we may have to take what limited data are to hand (so-called 'encountered data'; see, for example, Patil, 1991) which are difficult to analyse using formal methods. If we are to collect even limited data for our purposes we may need to abandon such hallowed principles as randomisation, not only in the search for access to rare events but also because we may thereby be able to use tailored methods which will yield high efficiency to counter the effects of expensive observation/measurement methods.

In this chapter we will consider a range of sampling methods that seek to overcome the rarity, sensitivity and cost problems *intrinsically*: that is to say, where their methodology is designed to aid concealment (and hence engender assurance) at the individual level or to yield 'super–efficient' population estimates to counter-balance high observation costs. Although strictly going beyond the range of 'finite population sample survey methods' they are very much part of the present day arsenal of sampling procedures and need to be represented in these pages.

We start with a review of a wide range of sampling principles for seeking rare population members with details of key methodological and applications-oriented references.

7.1   Some initial principles and some obstacles

We noted above that it may not be possible to collect sample data to a specified sampling design. Pollution 'hot spots' occur from time to time and at different locations on a landfill site for contaminated refuse. We are unlikely to be able to observe them at random. We may be lucky enough that measurements happen to have been made on the odd occasion when a location went 'hot'. Thus we 'encounter' a few observations. Such encountered data are not readily analysed statistically but may be all we can obtain (see Patil, 1991; Barnett, 2002). A more interpretable concept of encountered data arises in the context of *transect sampling*; see Section 8.2 below.

The pollution hot spot example illustrates two complicating features in data collection; matters of interest in the population may only occur rarely and we may have restricted control over what it is possible to observe.

For rare events *per se*, interests in identifying them in a population go back a long way. Two key references on sampling rare events are review papers by Cox and Snell (1979), who mention auditing as one relevant applications field, where incidence rates may be between 0.01 and 0.0001, and by Kalton and Anderson (1986).

Kalton and Anderson (1986) explain that when a list or sampling frame exists for the rare population members, then there are no special sampling difficulties and the standard finite sampling methods can be used. With no such list however (as is often the case) then special methods are needed to sample the rare population members. They consider incidence rates for the rare population members ranging from less than 1 in 1000 to as high as 1 in 10 (with special concern of health/medical surveys). They consider general **screening methods** for finding the rare members, where different incidence rates apply in different sub-populations, in which case **disproportionate sampling** may be relevant.

In relation to **screening**, Sudman (1972) remarks 'it may be efficient to screen a very large sample ... for a rare population, and to use the screening information for future samples'. This attitude embodies much of the thinking about sampling from populations where we want to find members with a rarely encountered characteristic. (Shorthand expressions such as 'sampling for a rare population' are widely used in spite of their semantic inexactitude!) The idea is that rare members by definition occur rarely in the population! So we try to use one or more concomitants (variables correlated with the characteristic of interest) to extract a sub-population in which the rare members are *more concentrated* and then sample from that sub-population.

Screening is not really a sampling procedure in its own right. It involves 'presampling' the population to make it easier to find the rare members. For example, in an epidemiological study to find individuals with a rare medical condition, known links with age, stature and previous medical complaints may be used to screen out a sub-population more likely to contain the targeted individuals. Screening may be conducted sequentially sometimes, either in terms of sequential application of progressively more sophisticated (and expensive) screening principles or in terms of progressively larger sections of the overall population. Sudman (1972) discusses this (and also refers to Bayesian methods for rare populations).

Screening methods (particularly in societal investigations) often employ concomitant information only in an informal way, reflected for example in the data collection process. A large cheaply assembled sample is first chosen to identify screening variables. These in turn are then used to screen further sample data. Often, we just sample clusters of the population members which can be expected to contain enhanced proportions of the rare members. See Ericksen (1976), Levy (1977), Kalton and Anderson (1986) (in addition to Sudman, 1972).

In **disproportionate sampling** we use stratified sampling with deliberate and structured departure from the proportional allocation approach. Suppose, for example, that we have specific information about the strata: namely that rare members of the population are more concentrated in some strata than in others; e.g. in certain geographic areas, or in certain sub-accounts of a financial record. In such cases, sampling the strata in which the rare members are concentrated at *higher sampling fractions* can lead to increased efficiency of estimation of population characteristics.

Other approaches to identification include **multiplicity** (or **network**) **sampling** methods where we obtain information on rare members from other members of the general population, or **snowball** (or **reputational**) **sampling** where encountered *rare members* help to identify others.

Lepkowski (1991) reviews probability sampling methods for the 'difficult-to-sample', again including use of disproportionate sampling, multiplicity sampling and multiple frames, where the notion of 'difficulty' may relate to *rarity* or to *cost*. Screening and multi-phase sampling methods are again considered, as are non-probability approaches such as *judgmental* or *purposive sampling*. Sudman, Sirken and Cowan (1988) discuss sampling 'rare and elusive populations', where costs of location are substantial. See also Sudman (1972); multiplicity network sampling is considered and (in the context of household surveys) *capture–recapture methods* (see Section 8.3 below). Hedges (1979) is concerned with sampling 'minority populations' including use of *disproportionate sampling*. A case study on disproportionate sampling of a rare population is described by Ericksen (1976); see also Mohlin, Pilley and Shaw (1991).

Multiplicity sampling was introduced mainly in social surveys to try to improve prospects of identifying rare members of a population by cross-reference from other rare members encountered in a sample. Thus sampled individuals with some rare characteristic in sampled households may be asked to direct us to others with that characteristic. Such **multiplicity sampling** (a form of **network sampling**) is discussed and illustrated in Kalton and Anderson (1986), referring to studies of cancer (Sirkin *et al.*, 1980), births and deaths (Nathan, 1976) and so on. Typically, a simple random sample of households might be cross-referred to other households through siblings, and individuals will have inclusion probabilities proportional to the number of linked households, and corresponding inverse weighting will be needed. Other institutional bases provide natural vehicles for multiplicity sampling: an example is use of hospital records in looking for patients with a rare disease.

The method was first suggested by Birnbaum and Sirken (1965) and has been substantially developed since then; see, for example, Levy (1977). Advantages include reduction in effort required to obtain a specific sample size of rare members (from the smaller set of cross-references); disadvantages include possible increase in nonresponse (reaction to having been 'fingered') and relative increase in sampling error due to the re-weighting (reducing to some extent the net advantage of the multiplicity design). Lepkowski (1991) further illustrates the uses of this method classifying different aspects of rarity that may prompt its use.

There are links between multiplicity sampling and **snowball sampling**. Both are forms of *network sampling* and use *reputational* or *chain-referral* principles. Snowball sampling focuses more exclusively on the rare members of the population, and progresses from one to another through several successive stages of referral. Its aim might be to establish a sampling frame of rare members from which a probability sample can be drawn, or the accumulated set of identified rare members may be taken to be the required 'sample' (but this is not strictly a probability sample).

Snowballing differs from other 'enhanced' methods we will discuss in its less formal employment of concomitant factors. It is more a principle than a method; an old principle at that. It seeks to exploit the fact that people like to 'gossip'! If we are trying to find an elusive person then we might 'ask around', and the closer we get, the more chance we have (usually) of finding the person. This is the basic idea behind snowballing, although 'person' and 'asking' need to be broadly interpreted: to find the elusive *item*, we *obtain information* on other items in such a way that we successively reduce the field in which we need to sample. See Berg (1988) on forms of snowball sampling.

We might also examine networks and links specifically and **network sampling** aims to identify links by tracing paths (*chains*) from one population unit to another. Lengths of chains, connectedness, and network density can all be of interest in different contexts. Fields of application have included sociological structure and medical epidemiology (e.g. drug addiction, police convictions and contagious diseases).

As the name suggests, paths are traced through series of referrals from one sampling unit to the next. The overall set constitutes the chain-referral sample or snowball sample. Thus an initial (random) sample leads to other *connected* units, all (or some) of which lead to others, and so on, yielding a set of chains. Various schemes use different numbers of stages (*waves*) with full or partial referral at each stage. Note that at any stage, a particular unit will refer only to another *connected* one.

Such an approach has been the subject of much statistical probabilistic analysis, starting with Goodman (1961, or even earlier in 1953). Some interesting applications include Daley and Kendall (1965; stochastic rumours), Cane (1966; epidemics and rumours), Gani and Jerwood (1971; chain binomial epidemic models), Berg (1983, random contact processes and snowball sampling). See also Biernacki and Waldorf (1981) who discuss identifying rehabilitated heroin addicts (see also Kaplan, Korf and Sterk, 1987) and Useem (1973) on failing to register for military service in the US.

Estimation of aggregate population characteristics from snowball samples (rather than of measures of relationship and chaining) is beset with some problems. These include the conditionality and frequent non-randomness of units appearing in successive waves, and non-response difficulties, which can lead to biased estimates whose precision is difficult to determine.

Sometimes the aim is to *assemble the sampling frame* of rare population members by fortuitously revealing a few of the rare members and using these to identify others *en route* to assembling the sampling frame, from which a probability sample can then be drawn. Alternatively, the successively identified rare members may be themselves adjudged to comprise the samples from which inferences are to be drawn—in the spirit of our earlier discussion of snowball sampling.

7.2   Composite sampling

We noted in Section 6.4.2 how the *randomised response* method can help to gain the confidence and maintain the anonymity of respondents in respect of the estimation of proportions in the population who are in certain 'sensitive' categories, e.g. drug-users or those with a criminal record. This approach was not specifically designed to handle rare events. Another method, which by its very nature can help to deal with *sensitive and rare* issues, is known as **composite sampling**.

One of the earliest applications of the method (Dorfman, 1943) was for the identification of individual US army personnel (and the estimation of the proportion in the army population) who were suffering from *Syphilis*. Suppose we have to screen a large population of size *N* either to identify all the individuals with *Syphilis* or to estimate the proportion with the disease. These are distinct problems since if we have identified all the individual sufferers we *know the proportion*, whilst if we know the proportion we *do not know who the individual sufferers are*. Composite sampling enables us to tackle either problem.

We will start with the problem of identification

We could take blood samples from each member of a group of *n* of the servicemen and *instead of testing each blood sample* we could mix the blood samples together to form a **pooled sample** (a little from each) and do just one test for *Syphilis* on the combined material. *If the test is negative we have cleared all* n *of the servicemen of Syphilis.*

This is the principle of composite sampling; that we have the ability to observe the presence of a condition by testing a composite sample over many individuals. It has two major advantages. Firstly, the composite result is not specific to any individuals—so is not intrusive if the presence or absence of the condition is a personal or sensitive issue. Secondly, if testing of the sample is an expensive or time-consuming matter, composite testing is inevitably cost-effective.

Such is the case at least if the test result is negative! Otherwise we must go further (with a positive composite result) and try to pinpoint the positive individuals. Thus it is always important to keep 'audit samples'—further sample material from each individual for later-stage testing. We return to the question of how to find the 'positives' a little later.

Not all testing situations are suitable for composite testing but many are and *composite sampling* is a widely researched and applied method. Areas of application include testing for disease, examining pollution (e.g. dangerous substances on contaminated land), assessing safety conditions (of bathing or drinking water, of physical structures, etc.), biodiversity, geostatistical measures, and so on. Specific examples are discussed in Watson (1936; insects carrying a plant virus, even earlier than Dorfman, 1943), Carter and Lowe (1986; forest floor properties), Green and Strawderman (1986; estimating timber volume), Baldock *et al.* (1990, sheep infection), Patil *et al.* (1992; various), Lovison *et al.* (1994; various), Barnett and Bown (2002, setting environmental standards).

Suppose then that our composite test yields a *positive result*. We will now need to go further to try to identify the specific positive individuals. The most obvious way

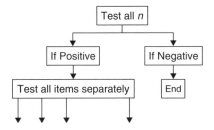

Fig. 7.1. Full retesting

forward is to test each of the individuals separately—what we might term *full retesting*, see Figure 7.1.

So we reach a situation where we have to conduct either just one test (if the composite is negative) or $n + 1$ tests if the composite is positive. But this can be highly economical. Suppose that the condition of interest occurs at random in a proportion p of the population. Then with probability

$$P(1) = (1 - p)^n \tag{7.1}$$

we will need just one test (all are negative) whilst with probability

$$P(n + 1) = 1 - (1 - p)^n \tag{7.2}$$

we will need $n + 1$ tests. So the *average number of tests needed* is $(n + 1) - n(1 - p)^n$.

The method is intended to be used when p is small (for a rare event). Thus if we have $p = 0.0005$ and $n = 20$ we find that on average just 1.2 tests are needed for each group of 20 individuals. So we have a highly economical approach which most of the time will not be probing into the condition of specific individuals.

There are other ways of seeking to identify the 'positives' when we encounter a positive composite result. Let us consider two or three of them. Figures 7.2, 7.3 and 7.4 show what might be termed *group retesting, cascading* and *'sudden death'*, which operate as follows.

In **group retesting** we divide the group of size n into k subgroups of sizes n_1, $n_2, \ldots, n_k$ if the first overall composite test is positive and then treat each of these subgroups as a second-stage composite sample. Each positive subgroup is then tested as in the full retesting approach and the process terminates at this third stage.

With **cascading**, we adopt a special and useful modification of group retesting where we operate hierarchically dividing each positive group or subgroup into *two equal parts* and continue composite testing on the successively smaller subgroups until all positives have been identified.

Finally, we might consider the **'sudden death'** approach where if the first composite test is positive, we treat individuals separately until we find a first 'positive' and then test the residue as a composite. If the residue group is negative we have finished. Otherwise, we test individuals until we find another 'positive' and then do a composite test on the new residual group, and so on.

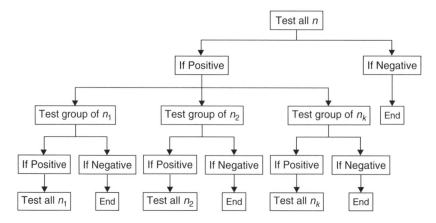

Fig. 7.2. Group retesting

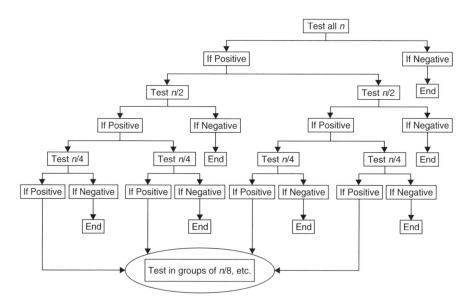

Fig. 7.3. Cascading

These methods have different propensities to yield a rapid resolution of the problem of identifying all positives depending on the values of n, of p and of subgroup sizes and of how the positives are distributed throughout the overall group. It is interesting to try out a range of possibilities in any practical situation.

Estimating the proportion p

This is another matter of great interest in composite sampling. Consider the situation where we take m random samples each of size n and carry out a composite test

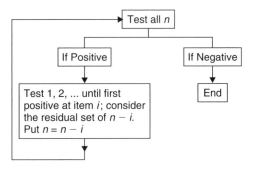

Fig. 7.4. 'Sudden death'

on each group of n. We do not seek to identify positives, but merely note if the sample gives a positive or negative composite result. (Note that as with *randomised response* in Section 6.4.2 we do not have to reveal individual circumstances.) Suppose we find that r out of the m are positive and $m - r$ are negative. Then r is just an observation from a binomial distribution $B(m, \phi)$ where ϕ is the probability that a random group gives a positive composite result. Thus we have the familiar result that r/m is the maximum likelihood estimator (MLE) of

$$\phi = 1 - (1 - p)^n \tag{7.3}$$

and by the transformation invariance property of the MLE we find the MLE of the population proportion p is just

$$p* = 1 - (1 - r/m)^{i/n}. \tag{7.4}$$

Boswell and Patil (1987) review the properties of this estimator in the context of earlier work by Rohde (1976), Elder *et al.* (1980), Garner *et al.* (1988) and others. It is not an entirely attractive estimator. It is biased and tends to overestimate p, but perhaps by no more than 10% in most situations (this is a conservative outcome: it will usually be better to overestimate p than to underestimate it).

Composite sampling has also been used when we are interested in a quantitative variable, which can in principle be measured on each individual but where we can measure its sum over the individuals in the composite sample. Only certain variables will have this summation property. Consider for example contamination of biological specimens in the form of a specific number of infected spores. The composite sample will show the sum of these. In such situations proposals have been made for testing for threshold levels, or estimating the mean and variance, of the variable. See, for example, Brumelle *et al.* (1984), Boswell and Patil (1987), Lancaster and McNulty (1998), Barnett and Bown (2002) and Barnett (2003).

Example 7.1

A random variable Y with mean μ and variance σ^2 is measured on individual members of a finite population. We can also measure Y on a (composite) set of n distinct members and its value will be the sum of the constituent values for the individual members. We take m composite samples with resulting Y values, $Y_1, Y_2, \ldots, Y_m$. Now, each Y_i has mean $n\mu$ and variance $n\sigma^2$.

Consider the estimator $\tilde{\mu} = \Sigma Y_i/(nm)$, which is clearly *unbiased* for μ since $E(\tilde{\mu}) = mn\mu/(nm) = \mu$. To estimate σ^2 we might consider basing it on the statistic $T = \Sigma(Y_i - n\tilde{\mu})^2$. Since each Y_i has mean $n\mu$ and variance $n\sigma^2$, $T/(m - 1)$ has expected value $n\sigma^2$. Hence $\Sigma(Y_i - \tilde{\mu})^2/[n(m - 1)]$ is *unbiased* for σ^2.

Clearly, we cannot estimate σ^2 if $m = 1$ (a single composite sample). For $m > 1$, we *can* estimate σ^2. For $m = 2$, an unbiased estimator of σ^2 is given by $(Y_1 - Y_2)^2/2n$.

7.3 Ranked set sampling

This is another method which goes back a long way (McIntyre, 1952, used the approach for estimating pasture yields in Australia) but it has had a major revitalisation of interest in recent years in view of the very high estimation efficiency yields it can achieve. Thus its role in this chapter is as the prime contender for estimating population characteristics when *sampling methods are complex and sampling costs are high.*

We noted an example of the use of ranked set sampling in the introductory chapter (in Section 1.4) where we considered the problem of estimating the mean pollution level in bathing water in a recreational lake (where the observation and measurement process is expensive). Rather than taking a simple random sample of, say, five observations and estimating the mean from the sample mean, in ranked set sampling we seek to 'spread out' our sample to obtain a better representation of the population. Thus we might choose five sites at random in the lake and seek expert advice to decide which of the five would be likely to give the largest pollution value; we then measure the pollution level at that site. We now take another random sample of five sites and with further expert guidance we measure what we expect to be the second largest pollution value. The process is continued until on the fifth occasion we seek to measure the lowest pollution level.

This sampling approach is known as **ranked set sampling**. We will explore some of its implications and properties. We will find that it can be a highly efficient sampling method, not only for estimating the mean but also percentiles, measures of spread, etc. or carrying out tests of significance, e.g. in regression modelling.

In respect of the 'ordering' process, we should note at this stage that

- the ordering need not be done on a subjective basis—we might order in terms of the values of a cheaply observed concomitant variable (e.g. weed cover)
- the resulting sample need not be *ordered*, i.e. with the first observation the largest, and so on
- even when the ordering process does not fully succeed we will still hope to achieve major efficiency gains over random sampling.

In summary, the objective is to exploit concomitant information (which might be cheaply and readily accessible, sometimes in subjective form) to seek to 'spread out' the sample values over their possible range and correspondingly to achieve a dramatic increase in efficiency over simple random sampling. There is an extensive literature on ranked set sampling since the early work by McIntyre (1952). See for example the reference lists in Barreto and Barnett (1999, 2001). Some of the milestones are as follows. Mathematical development of McIntyre's ideas was pursued by Takahasi and Wakimoto (1968). Dell and Clutter (1972) relaxed the earlier assumption of *perfect ordering* and showed the extent to which efficiency gains were still realisable as long as some correlation remained between the objective and concomitant variables. Stokes (1977, 1980) and Ridout and Cobby (1987) further developed the study of the effects of *imperfect ordering*. To this time most work concentrated on the use of the **ranked set sample mean** (the arithmetic average sample value). Sinha *et al.* (1996), Stokes (1995) and Barnett and Moore (1997) introduced optimal linear estimators, which could show even further improvement in efficiency of estimation of means and standard deviations, etc. Barnett (1999) explored design factors involving the use of different numbers of observations at the different rank orderings.

We will return to some of these considerations later but must start by setting out the mathematical form of ranked set sampling and exploring the properties of resulting estimators.

7.3.1 The ranked set sample and ranked set sample mean

Ranked set sampling operates as follows. We consider n conceptual random samples each of size n, of observations of the random variable Y:

$$y_{11}, y_{21} \ldots, y_{1n}$$
$$y_{21}, y_{22} \ldots, y_{2n}$$
$$\vdots$$
$$y_{n1}, y_{n2} \ldots, y_{nn}$$

From each conceptual sample we take one measured observation $y_{i(i)}$: the ith ordered value in the ith sample ($i = 1, 2, \ldots, n$). The ranked set sample is then defined as the set of observed values $y_{1(1)}, y_{2(2)}, \ldots, y_{n(n)}$. In the early applications reviewed above, the mean μ_Y of the underlying distribution was estimated by

$$\bar{\bar{y}} = \sum y_{i(i)} / n \tag{7.5}$$

which is the **ranked set sample mean**, which compares favourably (indeed impressively) in efficiency with $\bar{y}$ (the mean of a random sample of size n). Note that we compare with a sample of size n, not of size n^2, because measurement is assumed to be of overriding effort in obtaining the data compared with the ordering process. We say that the *ordering is perfect* if each $y_{i(i)}$ is an observation of the ith order statistic of a sample of size n.

We will find that $\bar{Y}$ is unbiased and it will turn out that (for $n > 2$) typically

$$\text{Var}(\bar{\bar{Y}}) < \text{Var}(\bar{Y})$$

often markedly so, for different sample sizes and for different distributions, if we have achieved *perfect ordering* in the sense described above.

When the ordering is not entirely correct, we will still achieve unbiasedness and increased efficiency compared with $\bar{Y}$ provided that there is non-zero correlation between the observed sample values and their claimed order. With zero correlation the ranking becomes equivalent to mere randomisation (and then we are no worse off than a random sample). We will obtain impressive efficiency gains, e.g., of 100% and more for a sample size $n = 5$ for both normal and exponential populations (see Dell and Clutter, 1972; Ridout and Cobby, 1987).

We need to show how these results come about. We will do so in the context of a random variable Y from a location/scale family; we assume that Y has distribution function (df) $F\{(y - \mu)/\sigma\}$. If Y is symmetric, μ is the mean, $E(Y)$; otherwise $E(Y)$ will be a linear combination of μ and σ. The *sample mean* $\bar{y} = \Sigma y_i/n$ is unbiased for $\mu_Y = E(Y)$ with variance σ_Y^2/n where $\sigma_Y^2 = \text{Var}(Y)$. This provides a useful reference against which to compare other estimators, e.g. the ranked set sample mean.

We can think of the ranked set sample values $y_{1(1)}, y_{2(2)}, \ldots, y_{n(n)}$ as the values of the *order statistics* $Y_{(1)}, Y_{(2)}, \ldots, Y_{(n)}$ in a sample of size n *but with a subtle difference*. The order statistics from a simple random sample will be correlated, whereas *the ranked set sample values are independent* (since they come from distinct potential samples). We can use properties of order statistics, simplified by this independence condition, to explore the behaviour of estimators based on the ranked set sample.

Let us start by examining $\bar{\bar{Y}}$. Consider the *reduced order statistics*

$$U_{(i)} = \left(Y_{(i)} - \mu\right)/\sigma$$

with means and variances denoted by α_i and v_i, respectively ($i = 1, 2, \ldots, n$). The $U_{(i)}$ are independent of μ and σ and

$$E\left(Y_{(i)}\right) = \mu + \sigma\alpha_i \qquad \text{Var}(Y_{(i)}) = \sigma^2 v_i. \tag{7.6}$$

Thus, in general,

$$E\left(\bar{\bar{Y}}\right) = \mu + \sigma\bar{\alpha} \tag{7.7}$$

and

$$\boxed{\text{Var}\left(\bar{\bar{Y}}\right) = \left(\Sigma v_i\right)\sigma^2/n^2.} \tag{7.8}$$

Clearly the sum of the ordered values in a sample is the same as the sum of the unordered values, i.e. $\Sigma Y_i = \Sigma Y_{(i)}$ so that taking expectations of each side we obtain

$$n\mu_Y = E\left(\Sigma Y_i\right) = E\left(\Sigma Y_{(i)}\right) = n\mu + \sigma\Sigma\alpha_i. \tag{7.9}$$

Thus

$$\mu_Y = \mu + \sigma\bar{\alpha} \tag{7.10}$$

which shows that $\bar{\bar{Y}}$ is unbiased for μ_Y. (If Y is symmetric, $\mu_Y = \mu$ so that we must have $\bar{\alpha} = 0$.)

Now consider the identity

$$\left[\sum(Y_i - \mu)\right]^2 = \left[\sum(Y_{(i)} - \mu)\right]^2.$$

Taking expectations of each side, noting that $E(Y) = \mu + \sigma\bar{\alpha}$ and that $E(Y_{(i)}) = \mu + \sigma\alpha_i$, we will find that

$$\mathrm{Var}(\bar{Y}) = \mathrm{Var}(\bar{\bar{Y}}) + \frac{\sigma^2}{n^2}\sum(\alpha_i - \bar{\alpha})^2. \tag{7.11}$$

The right-hand term $(\sigma^2/n^2)\sum(\alpha_i - \bar{\alpha})^2$ is essentially positive so we conclude, as claimed above, that

$$\mathrm{Var}(\bar{Y}) > \mathrm{Var}(\bar{\bar{Y}}).$$

The relative efficiency of $\bar{\bar{Y}}$ and $\bar{Y}$ is thus

$$e = \frac{\mathrm{Var}(\bar{Y})}{\mathrm{Var}(\bar{\bar{Y}})} = 1 + \frac{\sum(\alpha_i - \bar{\alpha})^2}{\sum v_i}.$$

For the special case where Y is symmetric we have $\mathrm{Var}(\bar{Y}) = \sigma^2/n$ and $\bar{\alpha} = 0$. So, from (7.8) and (7.11) we obtain

$$n = \sum v_i + \sum \alpha_i^2 \tag{7.12}$$

and the relative efficiency e can now be written just in terms of the α_i as

$$e = \left(1 - \sum \alpha_i^2 / n\right)^{-1}. \tag{7.13}$$

It can be shown that $1 < e \leqslant (n + 1)/2$ so that the potential efficiency gains from using the ranked set sample mean can be very high.

Example 7.2

Suppose we contemplate taking ranked set samples of sizes 3, 5 and 10 from normal, exponential and uniform distributions. What efficiency gains can we expect from estimating the population mean by the ranked set sample mean rather than by the mean of a random sample of the same size? The results are as follows.

	Normal	Exponential	Uniform
$n = 3$	1.91	1.64	2.00
$n = 5$	2.77	2.19	3.00
$n = 10$	4.79	3.41	5.00

We should note the magnitude of these efficiency gains! If e is of the order of 3 or 4 this is equivalent to the ranked set sample being as informative as a random sample which is 70% to 100% larger than the actual sample size. The sample sizes (3, 5, 10) chosen in this example might seem very small but this reflects the tendency in ranked set sampling to keep basic sample sizes small (less than, say, 10) in order to facilitate the ordering process. Larger sample sizes are achieved by replication. See, for example, Exercise 7.3 below. We also note that the upper limit to efficiency gains of the ranked set sample mean $((n + 1)/2$, shown below (7.13) above) is achieved for the uniform distribution.

7.3.2 Optimal ranked set sample estimators

We have so far considered estimating the mean of a population by the *ranked set sample mean* which is just the arithmetic average of the ranked set sample values, i.e. that linear combination which gives equal weight to each of the values. But why give equal weights? We could consider a more general linear combination: the *general linear estimator* (*L-estimator*) based on ordered values

$$\tilde{\mu} = \sum \gamma_i Y_{i(i)}$$

of which the ranked set sample mean $\bar{\bar{Y}} = \sum Y_{i(i)}/n$ is just a specific case.

Clearly the choice of $\sum \gamma_i = 1$ will ensure that $\tilde{\mu}$ is unbiased for μ_x. Subject to $\tilde{\mu}$ being unbiased we could choose the γ_i to minimize $\text{Var}(\tilde{\mu})$ rather than merely taking $\gamma_i = 1/n$ as for $\bar{\bar{Y}}$.

This prospect has been studied; see, for example, Stokes (1995), Sinha *et al.* (1996) and for a more comprehensive examination of optimality, Barnett and Moore (1997). The optimal estimator in the class of $\tilde{\mu}$ has been derived (Barnett and Moore, 1997) for the location/scale family of distributions as one of the two components of the jointly optimal pair $(\tilde{\mu}, \tilde{\sigma})$ where $\tilde{\sigma}$ is also in the class of linear order statistics estimators. The forms of the estimators are obtained from that special-case version of the general optimal order statistics estimators (David, 1981, Section 6.2 derived from Lloyd, 1952) which arises when we set the covariances (correlations) of the order statistics to zero (which we noted to be true for the ranked set sample).

The optimal (best linear unbiased) L-estimator of $\theta = (\mu, \sigma)$, which is known as the **ranked-set BLUE**, can thus be shown to take the form $\theta = (\mu^*, \sigma^*)$ where

$$\mu^* = \sum_{j=1}^{n} \gamma_i Y_{i(i)}, \quad \sigma^* = \sum_{j=1}^{n} \eta_i Y_{i(i)} \tag{7.14}$$

with

$$\gamma_i = \frac{(1/\nu_i)\left[\sum_{j=1}^{n}\left(\alpha_j^2/\nu_j\right) - \alpha_i \sum_{j=1}^{n}\left(\alpha_j/\nu_j\right)\right]}{\Delta},$$

$$\eta_i = \frac{(1/\nu_i)\left[\alpha_i \sum_{j=1}^{n}\left(1/\nu_j\right) - \sum_{j=1}^{n}\left(\alpha_j/\nu_j\right)\right]}{\Delta}$$

and

$$\Delta = \sum\left(\frac{\alpha_j^2}{\nu_j}\right)\sum\left(\frac{1}{\nu_j}\right) - \left[\sum\left(\frac{\alpha_j}{\nu_j}\right)\right]^2.$$

We can readily show that

$$\mathrm{Var}(\mu^*) = \sigma^2 \frac{\sum_{i=1}^{n}(\alpha_i^2/\nu_i)}{\Delta}$$

and

$$\mathrm{Var}(\sigma^*) = \sigma^2 \frac{\sum_{i=1}^{n}(1/\nu_i)}{\Delta}. \tag{7.15}$$

In the special case where *Y is symmetric* these results reduce to:

$$\mu^* = \frac{\sum(Y_{i(i)}/\nu_i)}{\sum(1/\nu_i)}$$

and

$$\sigma^* \frac{\sum(\alpha_i Y_{i(i)}/\nu_i)}{\sum(\alpha_i^2/\nu_i)} \tag{7.16}$$

with variances $\mathrm{Var}(\mu^*) = \sigma^2/\sum(1/\nu_i)$ and $\mathrm{Var}(\sigma^*) = \sigma^2/\sum(\alpha_i^2/\nu_i)$ (see also Sinha *et al.*, 1996; Barnett and Moore, 1997; and Barnett, 2003).

Recalling that $\mathrm{Var}(\bar{\bar{Y}}) = \sigma^2 \Sigma v_i/(n^2)$ we find, in the symmetric case, that the efficiency of μ^* relative to $\bar{\bar{Y}}$ takes the simple form

$$e(\bar{\bar{Y}}, \mu^*) = \frac{\mathrm{Var}(\bar{\bar{Y}})}{\mathrm{Var}(\mu^*)} = \frac{\left(\sum v_i\right)\sum (1/v_i)}{n^2} \tag{7.17}$$

The efficiency gain of μ^* over $\bar{\bar{Y}}$ is nothing like that of $\bar{\bar{Y}}$ over $\bar{Y}$. For example, we find efficiency gains of 4% and 9% at sample sizes of 5 and 10 for the normal distribution. For the exponential distribution the effects are more impressive with efficiency gains of 10% and 25% at sample sizes of 5 and 10, respectively. See Barnett and Moore (1997) for more details.

7.3.3 Imperfect ordering

We have so far assumed that the anticipated ordering in the ranked set sample is achieved, i.e. that $y_{i(i)}$ is a realisation of $Y_{(i)}$. However this is not necessarily true. In some problems, limitation of expert knowledge or a concomitant variable poorly correlated with the variable of interest may lead to imprecisions in the ordering of the sample values. Dell and Clutter (1972), however, have shown that in this case $\bar{\bar{Y}}$ will still be unbiased and will yield an efficiency gain over $\bar{Y}$ provided there is non-zero correlation between the sample values and the attributed order.

It is possible to quantify this effect for the location/scale family of distributions—see Barnett and Moore (1997). They assume that the ranking is achieved in terms of a concomitant variable X in order to identify the ith 'ordered' value in the ith conceptual sample (see also Stokes 1977, 1980). It is possible to determine the optimal ranked-set estimator (the **ranked set concomitant BLUE**) in this modified situation when Y and X have correlation coefficient ρ (Barnett and Moore, 1997). The optimal estimator μ_ρ^* is found to be unbiased with variance

$$\mathrm{Var}(\mu_\rho^*) = \sigma^2 \sum (\alpha_j^2/w_j)/\Delta'$$

where

$$\Delta' = \left(\sum (\alpha_j^2/w_j)\right)\sum (1/w_j) - \left(\sum (\alpha_j/w_j)\right)^2$$

and

$$w_i = (1 - \rho^2) + \rho^2 v_i.$$

Of course, if $\rho = 1$ or $\rho = -1$, we have $w_i = v_i$ and revert to the results above for perfect ordering. If $\rho = 0$ then $\mu_\rho^* \equiv \bar{Y}$ with variance σ^2/n and then no advantage accrues (and no loss arises) from use of the ranked set sample.

The rate of loss of efficiency gain for μ_ρ^* over $\bar{Y}$ can be quite rapid: e.g. for sample sizes of 5 and 10 in a normal distribution it falls from 4% and 9% to just 1% and

2% when ρ falls to just 0.9; for the exponential distribution gains of 10% and 25% (at $n = 5$ and $n = 10$) fall to 1.5% and 2.5% when ρ falls to 0.8.

The efficiency gains of $\overline{\overline{Y}}$ over $\overline{Y}$ are more robust and fall less dramatically with departure in the value of ρ from 1 (or -1).

7.4 Exercises

7.1 A survey is to be conducted to estimate characteristics of a population which contains a small proportion of members with a rare condition. Members know whether or not they have the condition, the presence of which can be detected from blood sample analysis, and support groups exist to advise and inform affected individuals.

By reference to an example of a condition of this sort, discuss the practicability, and relative advantages and disadvantages, of

- randomised response methods (Section 6.4.2)
- composite sampling
- snowballing

as methods of obtaining information on the incidence of the condition in the population.

7.2 The pollution level of a biological specimen is measured by a variable Y which is additive in its effect so that a set of n different specimens will have a cumulative (composite) pollution level $\sum y_i = n\bar{y}$ which is the sum of the separate levels, y_i.

It is important to detect whether any specimen in a set of size n has a pollution level as high as a control value y_0, in which case the set is *rejected*.

The 'rule of n' proposes rejection if $\bar{y} \geq y_0/n$.

Discuss the properties of this rule by considering the different prospects:

$$\text{(i)} \quad y_i = 0 \ (i \neq j); \ y_j = y_0$$

and

$$\text{(ii)} \quad y_i = y_2 = \cdots = y_n = y_0/n.$$

7.3 For a random variable Y with mean μ and variance σ^2, consider the conceptual samples

$$y_{11}, y_{21}, \ldots, y_{15}$$
$$y_{21}, y_{22}, \ldots, y_{25}$$
$$\vdots$$
$$y_{51}, y_{52}, \ldots, y_{55}$$

which might be used in drawing a *ranked set sample* of size 5. Instead of observing the ranked set sample $y_{1(1)}, y_{2(2)}, \ldots, y_{5(5)}$, suppose we observe the set of five *sample medians* $y_{1(3)}, y_{2(3)}, \ldots, y_{5(3)}$. Examine, for normal, double-exponential (with density proportional to $\exp(-|y|)$ and uniformly distributed Y, how the estimator $\bar{m} = \sum y_{i(3)}/5$ (which we might term the *memedian*) compares in bias and efficiency with the ranked set sample mean, $\sum y_{i(i)}/5$, for estimating μ.

8
Sampling natural phenomena

We often need to sample wildlife communities; we must collect appropriate data, to estimate density or abundance characteristics of the population or to aid decision-making, taking account of inaccessibility problems.

- It is necessary to seek to avoid bias and unrepresentativity which can occur (8.1) due to *size-biasing* or *length-biasing* or to use *weighting methods* to overcome such problems.
- *Transect sampling* methods (8.2) can be used to observe elusive population members and to estimate the extent of their presence.
- Capture–recapture or mark–recapture methods (8.3) can be used to estimate a wildlife population size, by catching and marking some members and examining how many of them are recaptured in a subsequent sample.

Another field in which it is important to be able to collect data, either to reflect characteristics of a population or to aid decision-making, is that of wildlife communities. They may consist of plants, trees in forests, fish stocks in fresh or marine water, birds or animals (in domesticity or in the wild). We can expect to encounter special access problems. Wild animals and birds are shy and may not reveal themselves, plants may be sparsely distributed and trees or fish stocks may be expendable (to weigh a tree or trawl for fish may mean that we have by sampling removed them from the very population we are trying to study).

These constraints may mean that it is difficult to obtain representative data and we will require special sampling methods, both *specifically*, in terms of how we seek to take a representative observation, and *generally*, in terms of what constitutes an acceptable sampling scheme to enable effective population inferences to be drawn.

For example, if we catch fish in a net we may lose smaller fish through the mesh (so that the catch is not representative; see Section 8.1 below) and we cannot readily arrange to draw successive individual fish at random (we will tend to catch them in *clusters* in the net). We will briefly examine different aspects of this problem including *quadrat sampling* and *weighting methods* (reflecting *specific* means of obtaining representative individuals from the population) *transect sampling* and *capture–recapture methods* (as *general* sampling systems for inference). *Adaptive sampling* (where the sampling method changes as we collect the data to reflect what we are learning about the locations of the target individuals) is discussed by Thompson and Seber (1996).

8.1 Difficulties in sampling plants and animals

In sampling plants or animals we may be concerned with estimating population *mean weight* or a relevant *proportion* (as in much of our earlier discussions) but frequently the objective is one of assessing *abundance*: how widespread or concentrated are the members of a species in the study area. Typically, we want to use statistical sampling to estimate species *density*; e.g. how commonly is the common tern encountered along the coast of Cumbria? However, we will usually be considering a finite geographic area over a specific time interval. So *density* is equivalent to population size and we note a change of emphasis from estimating means to estimating *population size*.

Let us start with the *specific* question of how we actually observe population members, say in environmental sampling particularly for ecological studies. If we wish to count the numbers of one, or of several, species of fish or bird or tree or plant, the basic need is to catch population members. This is often done in clusters rather than one by one (but see *transect sampling* in Section 8.2 below); for aquatic wildlife we might cast a net of given size into a pond, river or sea and count what it trawls. Corresponding to 'the net' for plant populations (to estimate population size or assess *biodiversity*) we might throw at random what is called a **quadrat** and do our count, or counts, within its boundary. A quadrat is usually a square (or round) frame of light wood or metal, commonly of one (or several) metre(s) side (or diameter). It is cast at random on to the study area and where it lands serves to define the search area in which we will observe all the individual plants in which we are interested. Quadrat sampling is used in many biological applications and with a variety of sampling designs including those we have studied above, such as *simple random sampling* or *stratified simple random sampling*. The quadrat (or its equivalent) is also used in many more structured sampling methods, e.g. *capture– (or mark–) recapture* (see Section 8.3 below), or *adaptive sampling*.

Consider the following problems:

- we are concerned with the weight distribution of fish in a pond and draw samples by catching them in a net
- we need to know (perhaps for health and safety reasons) the distribution of the lengths of fibre in a cotton crop and draw samples by scattering cotton fibres over a white tray and measuring the lengths of those crossing a randomly drawn line of prescribed length (so-called **line-intercept sampling**).

Both involve the 'quadrat' principle; based on the net for the fish and the white tray for the fibres. But they both present a similar difficulty, in that *neither set of observations will be representative of the true distribution*.

Why is this? The sample of fish will under-represent smaller fish, which will have fallen through the net. Likewise the sample of fibres will tend to include more longer ones that it should since the longer the fibre the higher the probability that it will cross the intercept line. Thus the true distribution and the observed distribution will tend to appear in relation to each other as the continuous and dotted lines respectively in Figure 8.1.

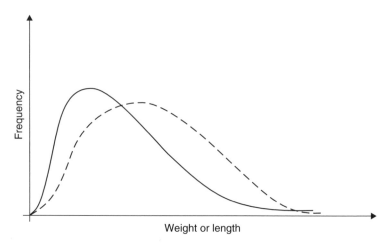

Fig. 8.1. Effects of size-biasing (— — — observed population; ——— true population)

So these sampling methods are likely to produce seriously biased results if we try to use them to estimate characteristics of the true population. They are what we call **size-biased** or **length-biased** samples. Any statistical inferences drawn from such samples must be inappropriate since they reflect on the distribution of *observed sizes,* not on the population of interest; viz. the distribution of *true sizes.* For example, estimates of the population mean or a specific quantile will tend to be too large (see Figure 8.1), perhaps markedly so.

Size-biasing is just one example of situations where the observed distribution does not correspond with the true distribution but is a *weighted version* of it. **Weighted distribution methods** are designed to restore the correspondence between the observed and true distributions so that valid inferences about the latter can be drawn. Much work has been done on this general theme over 60 years or more (e.g. see Fisher, 1934). However, specific concern for *size-biased* or *length-biased sampling* probably did not occur before the 1960s (Cox, 1952/1962; see Patil *et al.*, 1988) and it was Rao (1965) who introduced *weighted distribution methods* to help to remove the biasing effects. (See also Patil, 1997.)

Let us see how the bias arises with weighted distributions. Suppose Y is a non-negative rv with pdf $f_\theta(y)$, but an observation y does not inevitably appear in the observed sample. Its appearance is governed by a *relative retention distribution* with pdf $g(y)$ so that what we actually sample is a rv Y^* with a *weighted distribution* having pdf.

$$f_\theta^*(y) = kf_\theta(y)\, g(y).$$ (8.1)

In (8.1), k is a normalising constant satisfying $k^{-1} = \int f_\theta(y)g(y)\,dy$.

It is the special case where $g(y) \propto y$ that yields a weighted variable Y^* which follows what is called the *sized-biased* or *length-biased* distribution above; Y^* will now have pdf.

$$f_\theta^*(y) = y f_\theta(y)/\mu \tag{8.2}$$

where $\mu = E(Y)$.

In general, if we know $g(y)$ in any situation, we could seek to adjust for its effect. The size-biased case illustrates this but shows that it is not entirely straightforward. From (8.2) we have, for example,

$$E(Y^*) = \int y^2 f_\theta(y)\, dy/\mu = \int (y - \mu)^2 f_\theta(y)\, dy/\mu + 1/\mu$$
$$= \mu\left(1 + \sigma^2/\mu^2\right) \tag{8.3}$$

where $\sigma^2 = \mathrm{Var}(Y)$.

The effect of this can be quite severe since if we take an *observed* sample of size n, the sample mean $\bar{y}^*$ must be biased upward by a factor $(1 + \sigma^2/\mu^2)$! So if we had, for example, $\sigma^2/\mu^2 = 4$ (that is, a coefficient of variation of just 2) we would have a highly biased estimate of μ with expected value 5μ. Furthermore σ^2 and μ will be unknown, so there is no obvious way of seeking to correct for the bias (but see Barnett, 2003, Section 5.2 on the use of the *harmonic mean* in this context).

Example 8.1

Let us consider a particular case where Y has a Poisson distribution with mean μ: i.e. $Y \sim \mathbf{P}(\mu)$ and we obtain a size-biased sample. Here $\sigma^2 = \mu$ so that the expected value of the observed sample mean is now just $\mu + 1$ and we can obtain an unbiased estimate of μ as $\bar{y}^* - 1$. But this is a special case and we cannot in general handle size-biasing in such a simple manner. The simplification is more easily understood if we seek to obtain the distribution of Y^*. It will have probability function $p^*(y)$ proportional to $yp(y)$ with $p(y) = e^{-\mu}\mu^y/y!$ for $y = 1, 2, \dots$. So

$$p^*(y) \propto e^{-\mu}\mu^y/\mu(y-1)! = e^{-\mu}\mu^{y-1}/(y-1)!$$

and we see that $Y^* - 1$ has a Poisson distribution with mean μ.

Many other forms of weighted distribution have been studied and used. For example, in the fish example where we have a net with mesh size y_0 we do not really have size-biasing; the weight function is essentially a *truncation* in which

$$g(y) = \begin{cases} 0 & y \leqslant y_0 \\ 1 & y > y_0 \end{cases}.$$

With unknown θ (whatever the form of this parameter) this will not usually be easy to handle. The exception is if Y has an exponential distribution with mean $1/\theta$ when Y^* is also exponential but with its origin translated to y_0 (rather than being at zero).

See Patil and Rao (1977) (also Rao, 1965) for examples of weighted distribution methods over various situations other than *size-biasing* or *truncation*.

8.2 Transect sampling

What is the most straightforward way of sampling mushrooms in a marsh or wood-peckers in a wood? Presumably we would walk over the marsh or through the wood and just look for them. This, in a formalised form to ensure randomness of encounter, is precisely what is often done. It is the basis of the **transect sampling** method adopted for and widely used in many biological and environmental applications: specifically to estimate the numbers of a species of animal or plant distributed over a geographic region.

Its most direct form is **line-transect sampling**. We draw a line at random across the region we wish to represent and seek the objects of interest (mushrooms or wood-peckers) by moving along the line and recording those that we observe as we go from one end of the line to the other. We may traverse several lines and accumulate our sample data from all of them.

The aim of such a method is to estimate the density or abundance of plants or fish or animals—a remarkably difficult and time-demanding activity. *Line-transect sampling* is likely to be as reasonable a method for this as can be obtained in most types of application although we will encounter many difficulties which conspire to impede its effectiveness and efficiency. Often the search is limited to a *strip* of equal width either side of the line and the approach is then known as **strip-census methods** (Hayne, 1949).

The statistical analysis of sample data from transect sampling is based on many simplifying assumptions about the contact and behavioural mechanisms of the real-life objects we study; see, for example, Eberhardt (1978) for a detailed discussion of this. Examples include initial assumptions that the search objects are not moving, are equally and independently liable to be observed (usually), are seen on only one occasion, are observed *at right angles* to the transect line (usually) and are unaffected (e.g. neither repelled, nor attracted) by the observation process. Of course, such assumptions may not hold in a given situation but the basic transect sampling methods provide a reasonable starting point for sampling and estimation. Where any such assumptions are patently unreasonable more complex sampling approaches have been developed; we will consider some examples of these as we proceed.

Some key references to transect sampling and related methods include the following. Seber (1982, 1986, 1992) give wide-ranging surveys of inference methods for animal abundance. See also Burnham *et al.* (1980) including density estimation for biological applications and Quang (1991). Ramsey *et al.* (1988) includes marine applications.

Further reviews and references to work on line-transect methods are included in Thompson (1992) and Buckland *et al.* (1993) which presents detailed methods on distance sampling for animal abundance.

There is an alternative approach where we take observations *in all directions* from a single point perhaps spending some time doing so to note all the individuals (birds or animals perhaps) in the vicinity. This is referred to as a **variable circular plot** or a **point-transect sample**. Perhaps several points are used and the data accumulated. This method is particularly useful for birds and for elusive animals: reviews include Ramsey and Scott (1979), Burnham *et al.* (1980), Buckland (1987) and Quang (1993). Other variations include allowing for the prospect that some individuals (perhaps 'larger' ones) may be more visible than smaller ones, at any specific distance (see, for example, Chen, 1996), or that we sight the individuals 'ahead', i.e. not just at right angles to the transect line (see, for example, Burnham and Anderson,1976).

We should note that transect sampling operates on many different scales, from microscopic inspection of plant products to study of fish or animals by means of aerial surveys of spawning beds or game parks with transect lines (search lines) many tens or hundreds of miles in length.

Transect sampling methods cover a wide range of situations and the methodology is extensive. We will not be able to represent this complexity here but will just examine some of the initial principles and methods. Wider coverage is given by Barnett (2003).

8.2.1 Strip transects: the simplest case

We start with the basic case where a line transect of length l is chosen at random over an observation region which contains individuals of some object of interest distributed over the region at random at *rate* λ (or with *density* δ). It is assumed (at this stage) that any such individuals, which lie within a strip of width w either side of the transect line, will be observed and that none outside the strip will be.

So if we observe n objects, we can estimate the density δ by

$$\tilde{\delta} = \frac{n}{2wl}. \tag{8.4}$$

If the observation region has area A we can correspondingly obtain an estimate of the total number of individuals in the region (the 'population size' for the strip) as

$$\tilde{T} = \frac{An}{2wl}. \tag{8.5}$$

The problem with this approach is its artificiality. It is unlikely in practice that there will be a distinct cut-off distance up to which *everything is observed* and beyond which *nothing is observed*. So the basic approach is just an approximation to what really happens and is also inevitably inefficient (in assuming that no individuals are observed outside the strip or in ignoring any that are sighted). And even as an approximation we have the problem of deciding what strip width $2w$ to assign. One possibility is to consider *all sightings* and to decide on $2w$ *post hoc* from the histogram of numbers of observations at different distances. If this has a distinct cut-off we obtain a conclusive choice for w but this is not common.

Example 8.2

Suppose that in a particular line-transect survey with a transect line of length 200 m, we observe 70 animals at distances as indicated in the histogram in Figure 8.2.

Clearly the sightings do not totally disappear beyond some distance w_0 but detectabilty beyond 20 m is much lower than it is up to 20 m. So it

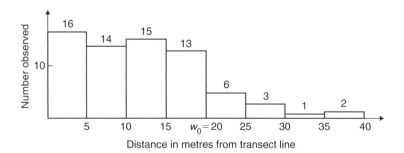

Fig. 8.2. Histogram of observations

might be reasonable to act as if we had a strip of width $2w = 2w_0 = 40$ m and to use the observations within the distance w_0 of the line as if they arose from the 'strip' with no other sightings being made. We could then estimate the density from (8.4) above as $58/8000 = 0.00725/\text{m}^2$. If the study region were 200 m long and 60 (or 200) m wide we would estimate the total population as 87 (or 290) animals in all. Note that we have had to discard the observations we made of 12 of the animals, which must be an inefficient practice.

8.2.2   Taking account of differential detectability

It is possible to proceed more formally as shown, for instance, by Burnham and Anderson (1976). We need to state the assumptions under which we will operate. Again, we assume

- that individuals are placed at random at rate λ over a region of area A
- that we take observations perpendicularly on either side of a randomly placed transect line of length l
- and, now, that whether or not we observe an object which is present is governed by a **detectability function**

$$g(x) = P \,(\text{object observed} \mid \text{object is at distance } x)$$

with $g(0) = 1$ (so that all individuals on the line will be observed).

Again we start by considering a notional strip of width w on either side of the line. Any individual in the strip will have a perpendicular distance x from the line, where x is an observation of a random variable, which we will assume is uniformly distributed on $(0, w)$.

This implies that there is some distance w from the line beyond which objects are *not to be found*, even though they are uniformly distributed up to this distance. Such a value w is purely notional to aid the mathematical development and does not need to be specified. In practice, it either does not enter explicitly into the results or we will only be concerned with what happens as w increases indefinitely.

This will imply that the number of individuals, n_w, in the strip is an observation of a random variable N_w with mean $E(N_w) = 2wl\delta$.

But all n_w individuals in the strip will not necessarily be observed. Whether we observe an individual *in the strip* will be tempered by the detection function; it will be observed with probability

$$P(\text{observe individual which is in strip}) = P_w = \int_0^w \frac{g(x)}{w}\,dx = E[g(X)]. \qquad (8.6)$$

Thus if we define $\mu_w = \int_0^w g(x)\,dx$, we have $P_w = \mu_w/w$.

Let us now consider those individuals we *actually observe*. Suppose there are n of them at distances $y_1, y_2, \ldots, y_n$; these are the X values of the observed individuals. There are other individuals in the strip but they are not observed. Thus the random variable N *describing the number of observed individuals* (of which we have the observed value $N = n$) will have expected value

$$E(N) = E_{N_w}\{E(N/N_w)\} = E_{N_w}\{P_w N_w\} = 2wl\delta P_w = 2l\delta\mu_w. \qquad (8.7)$$

and we can estimate δ as

$$\tilde{\delta} = \frac{n\tilde{m}_w}{2l} \qquad (8.8)$$

where $\tilde{m}_w$ is an estimate μ_w^{-1}.

It happens that we can use the distances of the observed individuals $y_1, y_2, \ldots, y_n$ to estimate μ_w^{-1}.

The pdf $f(y)$ of Y turns out to be $g(y)/\mu_w$ $(0 \leqslant y \leqslant w)$ with $g(0) = 1$. So we find that

$$f(0) = \frac{1}{\mu_w} \qquad (8.9)$$

and using (8.6) we can estimate δ as

$$\tilde{\delta} = \frac{n\tilde{f}(0)}{2l}. \qquad (8.10)$$

where $\tilde{f}(0)$ is an estimate of $f(0)$.

Thus we have only to estimate $f(y)$, or specifically $f(0)$, to estimate δ. Comparing (8.10) with (8.4), it is as if there was a strip of width $1/f(0)$ in which *all individuals present were observed*. This is sometimes known as the **equivalent strip**. So to proceed further we need to be able to estimate $f(y)$, or specifically $f(0)$, in the context of reasonable forms that might be adopted for the detectability function $g(x)$.

There has been much work done on this and we can briefly consider two possible approaches. We need to start by trying to fit some appropriate parametric family of distributions $f_\theta(y)$ to the observations $y_1, y_2, \ldots, y_n$ and then to incorporate an appropriate form of detectability function $g(x)$.

1. The simplest case would be to assume that *all* individuals which are present will be observed up to a distance w, and *none* beyond that distance, so that

$$g(x) = 1 \ (0 \leqslant x \leqslant 1).$$

Here $f(y) = 1/w \ (0 \leqslant y \leqslant w)$ also, and we would have $f(0) = w^{-1}$ which, from (8.10) gives

$$\tilde{\delta} = \frac{n}{2wl}.$$

This is just the 'simplest case' prospect of (8.4) which we obtained in Section 8.2.1.

2. A more sophisticated approach would be to try use the observed distances $y_1, y_2, \ldots, y_n$ to estimate the distribution of Y and then to use $\tilde{f}(0)$ to estimate δ from (8.10). We could, for example, use the grouped relative frequency distribution of our observations (cf. Example 8.2) and then fit a distribution to it (if n were large enough) perhaps checking the fit with a χ^2 goodness-of-fit test.

There are even more advanced density estimation methods that could be used. These prospects have been widely discussed. For example, see Gates (1979); Eberhardt (1978); Pollock (1978); Burnham *et al.* (1980); Seber (1982), Buckland and Turnock (1992); Buckland *et al.* (1993) and Chen (1996). See also Burnham and Anderson (1976) who use isotonic regression methods.

A well-illustrated lay review of sampling wildlife populations is given by Manly and McDonald (1996); this includes transect sampling and capture–recapture methods (the topic of the following section).

8.3 Capture–recapture methods

We now consider another widely used and complex set of sampling and estimation methods used for inferring total population size. These are based on the principle of sampling at random a subset of the population, 'marking' those that are 'captured' and returning them to the population. A subsequent independent random subset is drawn and the number of marked individuals 'recaptured' is noted. The combined data consisting of the two sample sizes and of the number marked in the second sample can then be used to estimate (principally) the overall *population* size. Such methods are known as **capture–recapture** or **mark–recapture** methods.

Let us consider an example. We wish to estimate the number N of fish in a pond. A random sample of n fish is caught and each of them is distinctively marked. They are released back into the pond. Subsequently, an independent random sample of m fish is caught and r of them are observed to have the markings. How can we estimate the population size N?

8.3.1 The estimators of Petersen and Chapman

We start with the most basic assumptions of the simplest approaches where the initial ('captured') random sample is 'marked' in a way which enables them to be unequivocally identified as having been chosen at the earlier stage. The capture probabilities and population structure are assumed to remain the same throughout (the population is 'closed'). Assuming such a homogeneous population is vital to justify the simple approaches to capture–recapture methodology.

What forms could 'marking' take? There are many possibilities, from rings on legs of birds, small radio transmitters under the skin of animals, radioisotope marking, cuts on fins of fish, colouring or dying of skin or fur, patterns cut in the bark of trees, coloured plastic ties, etc. Seber (1986), for example, discusses the use of small radio transmitters, remote heat-sensing and underwater acoustics, etc.

Let us consider this approach in some detail. Suppose an initial random sample of size n is chosen from a population of size N. Each of its members is 'marked' and they are returned to the population. A second random sample of size m is found to contain r marked individuals. How can we estimate N?

If the samples are chosen at random *without replacement* (as they would be if the fish were caught in a net, for instance) and there are no reasons to expect any distortions (such as 'trap shyness') to make the sample unrepresentative, then the probability $P(r)$ of obtaining r marked individuals in the second sample is given by the *hypergeometric distribution* as

$$P(r) = \frac{\binom{N-n}{m-r}\binom{n}{r}}{\binom{N}{m}}. \tag{8.11}$$

But if N is large in relation to n (as is likely in real-life applications) we can use instead either the *binomial* or the *Poisson* approximations to the *hypergeometric* distribution. These would give:

$$\text{Binominal:} \qquad P(r) \sim \binom{m}{r}\left(\frac{n}{N}\right)^r \left(1 - \frac{n}{N}\right)^{m-r}. \tag{8.12}$$

$$\text{Poisson:} \qquad P(r) \sim \frac{e^{\frac{-mn}{N}}}{(r)!}\left(\frac{mn}{N}\right)^r. \tag{8.13}$$

These distributions both have means mn/N. So the problem we now face is, for known n and m, estimating N as (*the reciprocal of*) the mean of the underlying hypergeometric, binomial, or Poisson distributions.

Equivalently, we would expect the proportions of marked fish to remain the same in both samples, i.e.

$$\frac{r}{m} \sim \frac{n}{N}.$$

This gives, on inversion, what is called the **Petersen estimator** of population size in the form

$$\tilde{N} = \frac{mn}{r}. \tag{8.14}$$

Let us examine the approach in terms of the binomial approximation. We assume r is an observation from the binomial distribution $B(m, n/N)$ and we estimate the binomial probability $p = n/N$ by $\hat{p} = r/m$, the *maximum likelihood estimator*. So, the estimator

$\tilde{N} = n/\hat{p}$ of $N = n/p$ is also the maximum likelihood estimator. We can employ Taylor expansion techniques to approximate the sampling properties of $1/\hat{p}$. These give a first-order approximation for an estimate of the sampling variance of $\tilde{N}$ in the form

$$\mathrm{Var}(\tilde{N}) \sim \frac{nm(n-r)(m-r)}{r^3}. \tag{8.15}$$

However the sampling distribution of $\tilde{N}$ tends to be highly skew, making $\tilde{N}$ seriously biased for N, for many values of m and n. Specifically, the bias can be estimated as

$$E(\tilde{N}) = N\left(1 + \frac{m-r}{r^2}\right).$$

Thus (8.15), as an approximate form for the variance of $\tilde{N}$, has limited value. A further complication arises if $r = 0$ (we obtain *no marked members*) since then $\tilde{N}$ has infinite estimated variance!

It is natural to try to obtain alternative estimators of N with lower bias and to seek approximately unbiased estimators of their variances (see, for example, Seber, 1982). One proposal is what is known as the **Chapman estimator**. This is obtained from the Petersen estimator by increasing each of r, m and n by 1 to yield the estimator

$$\hat{N} = \frac{(n+1)(m+1)}{(r+1)} - 1 \tag{8.16}$$

which is found to be unbiased if $n + m > N$ (and approximately so otherwise). This has an approximately unbiased estimator of its variance in the form

$$\mathrm{Var}(\hat{N}) = \frac{(n+1)(m+1)(n-r)(m-r)}{(r+1)^2(r+2)} \tag{8.17}$$

which is finite even if $r = 0$.

Example 8.3

A field study of small mammals in woodland involved capture, marking and release of 280 animals. Later a random catch of 210 of such animals contained 55 of the previously marked ones. We can estimate as $\tilde{N}$ the number N of small mammals in the woodland by the Petersen or the Chapman estimators, giving:

$$\text{Petersen}\quad \tilde{N} = \frac{280 \times 210}{55} = 1069$$

$$\text{Chapman}\quad \tilde{N} = \frac{281 \times 211}{56} - 1 = 1058.$$

The estimated variances are respectively 12 325 and 11 568 so that approximate 95% confidence intervals for N are obtained as $851 < N < 1287$ and $847 < N < 1269$, respectively.

We will not be able to further develop the extensive capture–recapture methodology in detail here. This is more widely covered in various texts and review papers: see, for example Cormack (1979), Seber (1986, 1992), Pollock (1991), Thompson (1992) and Barnett (2003). A well-illustrated lay review of sampling wild-life populations is given by Manly and McDonald (1996).

However, in conclusion, we should examine some of the assumptions underlying capture–recapture which, if not satisfied, will make it necessary to employ more sophisticated methods. Firstly, we need to *capture* a sample of animals or insects (capture–recapture) or *mark* some plants (mark–recapture). Various complications are possible here. For example:

- The marking process may be multi-stage.
- The individuals in the samples may be chosen *with* or *without replacement*.
- Some individual population members may tend to *avoid*, or to *seek*, capture, being '*trap shy*' or even '*trap happy*'.
- Sometimes the catching or marking process can intrinsically affect the population. In extreme cases, plants or animals may even be destroyed by the intervention process.
- This makes the sample depart from randomness and the population will change in form from capture to recapture (rather than remaining '*closed*').
- Further, the population may markedly change from capture to recapture due to *births*, *deaths*, *immigration* or *emigration*.

The simplest approaches to capture–recapture such as those described above will need to assume randomness of the samples with constant capture probabilities in a fixed ('closed') population (no births, deaths, etc.) without capture-related effects (of being 'trap shy' or 'trap happy' or marks being lost). It is not straightforward to ensure that these conditions hold.

If they are not satisfied we will need to adopt methods designed for '*open population models*' where the population can change between 'captures' or be influenced by the catching, marking and recapture processes; see Seber (1986, 1992) and the other references mentioned earlier for details on such approaches.

8.4 Exercises

8.1 We have initially marked a number of individuals in a region and we then make k passes through the region (e.g. on transect lines), on each of which we take random recapture samples.

The k recapture samples have m_i observed animals, of which r_i turn out to be marked, in sample i ($i = 1, 2, ..., k$). Suppose n were initially marked, out of a population of N. So we have $m = \sum m_i$ recaptured individuals in all of which $r = \sum r_i$ are marked.

Show how an estimator $\tilde{N}$ of population size N based on the Petersen form can be reinterpreted as a classical *finite population ratio estimate* as discussed in Chapter 3.

From the results for ratio estimates, show that $\tilde{N}$ is approximately unbiased and obtain an estimate of its variance. *Note that* if the m_i and r_i are roughly proportional (e.g. as is likely to arise with an even spread of marked animals throughout the region of study), then $\tilde{N}$ will be a highly efficient estimator.

8.2   A transect line of length 400 m is drawn through a game reserve and the line transect survey observes 80 pheasants. The distances from the line (in m to the nearest m) yield the following frequency distribution.

Distance	≤5	6–10	11–15	16–20	21–25	26–30	31–35	36–40
No of pheasants	21	17	14	10	9	4	4	1

The reserve is of area 12 hectares ($120\,000\text{m}^2$). Estimate the number of pheasants in the reserve taking account of possible varying detectability with distance from the transect line.

Numerical solutions to exercises

Exercise 2.1
$3.106 < \bar{Y} < 3.404$

Exercise 2.2
Data in *Exercise* $y_T = 40\,140$ days   Estimated standard error $= 1037$
Data in *Example 2.3* $y_T = 46\,656$ days   Estimated standard error $= 1738$

Exercise 2.3
$y_T = 4749$ books   Confidence interval $4464 < Y_T < 5035$
Need 78 shelves

Exercise 2.5
Need sample of 1667 observations

Exercise 3.1
Ratio estimator $y_{RT} = 422.5$ Estimated standard error $= 138.2$
Regression estimator $y_{LT} = 422.5$ Estimated standard error $= 137.9$
Sample total estimate $y_T\ = 445.8$ Estimated standard error $= 680.5$
Estimated efficiency of ratio or regression estimator relative to sample total: 20.3%

Exercise 3.2
Estimated efficiency of ratio estimator relative to sample total: 205%

Need nine units or more

Exercise 3.3
$\bar{y}_L = 3.38$, with estimated standard error $= 0.119$

Exercise 4.1
Stratum sample sizes: 13, 30, 7

Exercise 4.2
$\bar{Y} = 8.438$ $S^2 = 4.313$
Proportional allocation: 25, 21, 16, 9, 10
Neyman allocation: 23, 24, 13, 10, 10

Estimated efficiency for sr sample mean relative to proportional allocation: 36.7%
Estimated efficiency for equal allocation relative to Neyman allocation: 36.1%

Exercise 4.3
Equal allocation: 40, 40, 40
Proportional allocation: 36, 72, 12
Neyman allocation: 38, 72, 10
Estimated efficiency for equal allocation relative to proportional allocation: 72.2%
Estimated efficiency for equal allocation relative to Neyman allocation: 72.0%

Exercise 4.5
Proportional allocation: 19, 22, 21, 17, 9, 5, 7
Neyman allocation: 9, 18, 18, 19, 13, 8, 15
Estimated efficiency for proportional allocation relative to Neyman allocation: 84%

Exercise 5.3
Estimated standard error of $\bar{y}_{c(b)} = 6.93$
Estimated standard error of $\bar{y}$ $= 0.449$

Exercise 7.3
All unbiased. Efficiency of ranked set sample mean relative to *memedian*:
   Normal 60%
   Double-exponential 41%
   Uniform 128%

Exercise 8.2
Estimated total number of pheasants:
- using fixed strip of width 51 m   418
- fitting an exponential detectability function   919

Appendix

Table 1 Random digits

51 74 23 99 67	61 32 28 69 84	94 62 67 86 24	98 33 41 19 95
63 38 06 86 54	99 00 65 26 94	02 82 90 23 07	79 62 67 80 60
35 30 58 21 46	06 72 17 10 94	25 21 31 75 96	49 28 24 00 49
63 43 36 82 69	65 51 18 37 88	61 38 44 12 45	32 92 85 88 65
98 25 37 55 26	01 91 82 81 46	74 71 12 94 97	24 02 71 37 07
02 63 21 17 69	71 50 80 89 56	38 15 70 11 48	43 40 45 86 98
64 55 22 21 82	48 22 28 06 00	61 54 13 43 91	82 78 12 23 29
85 07 26 13 89	01 10 07 82 04	59 63 69 36 03	69 11 15 83 80
58 54 16 24 15	51 54 44 82 00	62 61 65 04 69	38 18 65 18 97
34 85 27 84 87	61 48 64 56 26	90 18 48 13 26	37 70 15 42 57
03 92 18 27 46	57 99 16 96 56	30 33 72 85 22	84 64 38 56 98
62 95 30 27 59	37 75 41 66 48	86 97 80 61 45	23 53 04 01 63
08 45 93 15 22	60 21 75 46 91	98 77 27 85 42	28 88 61 08 84
07 08 55 18 40	45 44 75 13 90	24 94 96 61 02	57 55 66 83 15
01 85 89 95 66	51 10 19 34 88	15 84 97 19 75	12 76 39 43 78
72 84 71 14 35	19 11 58 49 26	50 11 17 17 76	86 31 57 20 18
88 78 28 16 84	13 52 53 94 53	75 45 69 30 96	73 89 65 70 31
45 17 75 65 57	28 40 19 72 12	25 12 74 75 67	60 40 60 81 19
96 76 28 12 54	22 01 11 94 25	71 96 16 16 88	68 64 36 74 45
43 31 67 72 30	24 02 94 08 63	38 32 36 66 02	69 36 38 25 39
50 44 66 44 21	66 06 58 05 62	68 15 54 35 02	42 35 48 96 32
22 66 22 15 86	26 63 75 41 99	58 42 36 72 24	58 37 52 18 51
96 24 40 14 51	23 22 30 88 57	95 67 47 29 83	94 69 40 06 07
31 73 91 61 19	60 20 72 93 48	98 57 07 23 69	65 95 39 69 58
78 60 73 99 84	43 89 94 36 45	56 69 47 07 41	90 22 91 07 12
84 37 90 61 56	70 10 23 98 05	85 11 34 76 60	76 48 45 34 60
36 67 10 08 23	98 93 35 08 86	99 29 76 29 81	33 34 91 58 93
07 28 59 07 48	89 64 58 89 75	83 85 62 27 89	30 14 78 56 27
10 15 83 87 60	79 24 31 66 56	21 48 24 06 93	91 98 94 05 49
55 19 68 97 65	03 73 52 16 56	00 53 55 90 27	33 42 29 38 87
53 81 29 13 39	35 01 20 71 34	62 33 74 82 14	53 73 19 09 03
51 86 32 68 92	33 98 74 66 99	40 14 71 94 58	45 94 19 38 81
35 91 70 29 13	80 03 54 07 27	96 94 78 32 66	50 95 52 74 33
37 71 67 95 13	20 02 44 95 94	64 85 04 05 72	01 32 90 76 14
93 66 13 83 27	92 79 64 64 72	28 54 96 53 84	48 14 52 98 94
02 96 08 45 65	13 05 00 41 84	93 07 54 72 59	21 45 57 09 77
49 83 43 48 35	82 88 33 69 96	72 36 04 19 76	47 45 15 18 60
84 60 71 62 46	40 80 81 30 37	34 39 23 05 38	25 15 35 71 30
18 17 30 88 71	44 91 14 88 47	89 23 30 63 15	56 34 20 47 89
79 69 10 61 78	71 32 76 95 62	87 00 22 58 40	92 54 01 75 25

This table is taken from Fisher and Yates: *Statistical Tables for Biological, Agricultural and Medical Research*, published by Longman Group Ltd., London (previously published by Oliver & Boyd, Edinburgh), and by permission of the authors and publishers.

Table 2 Double-tailed percentage points, $t_\nu(\alpha)$ and z_α, for Student's t-distribution, and the normal distribution.

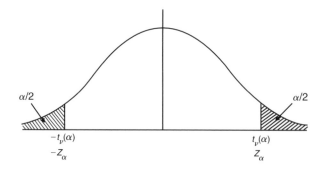

Degrees of freedom	α					
	0.10	0.05	0.02	0.01	0.002	0.001
t-Distribution						
ν						
1	6.314	12.71	31.82	63.66	318.3	636.6
2	2.920	4.303	6.965	9.925	22.33	31.60
3	2.353	3.182	4.541	5.841	10.22	12.94
4	2.132	2.776	3.747	4.604	7.173	8.610
5	2.015	2.571	3.365	4.032	5.893	6.859
6	1.943	2.447	3.143	3.707	5.208	5.959
7	1.895	2.365	2.998	3.499	4.785	5.405
8	1.860	2.306	2.896	3.355	4.501	5.041
9	1.833	2.262	2.821	3.250	4.297	4.781
10	1.812	2.228	2.764	3.169	4.144	4.587
12	1.782	2.179	2.681	3.055	3.930	4.318
14	1.761	2.145	2.624	2.977	3.787	4.140
16	1.746	2.120	2.583	2.921	3.686	4.015
18	1.734	2.101	2.552	2.878	3.611	3.922
20	1.725	2.086	2.528	2.845	3.552	3.850
25	1.708	2.060	2.485	2.787	3.450	3.725
30	1.697	2.042	2.457	2.750	3.385	3.646
40	1.684	2.021	2.423	2.704	3.307	3.551
60	1.671	2.000	2.390	2.660	3.232	3.460
80	1.664	1.990	2.374	2.639	3.195	3.415
∞	1.645	1.960	2.326	2.576	3.090	3.291
Normal distribution						
z_α	1.645	1.960	2.326	2.576	3.090	3.291

Bibliography and references

Note: Numbers in parentheses following each entry indicate the pages on which the publication is referenced.

Anthony, Y. C. K. (1983). Double bootstrap estimation of variance under systematic sampling with probability proportional to size. *J. Statist. Comp. Sim.*, **31**(2). (145)

Arens, A. A. and Loebbecke, J. K. (1981). *Applications of Statistical Sampling to Auditing*. Prentice-Hall, New York. (8)

Assael, M. and Keon, J. (1982). Non-sampling vs sampling errors in survey research. *J. Marketing*, **46**, 114–123. (160)

Baldock, F. C., Lyndal Murphy, M. and Pearse, B. (1990). An assessment of a composite sampling method for counting strongyle eggs in sheep faeces. *Austral. Vet. J.*, **67**, 165–169. (194)

Barnett, V. (1999). Ranked set sample design for environmental investigations. *J. Environ. Ecol. Statist.*, **6**, 59–74. (199)

Barnett, V. (2002). Encountered data. In El-Shaarawi, A. and Piegorsch, W. (Eds) (2002). *Encyclopaedia of Environmetrics, Vol. 2*. Wiley, Chichester, pp 668–669. (168, 191)

Barnett, V. (2003). *Environmental Statistics*. Wiley, New York (in press). (26, 190, 197, 203, 209, 211, 217)

Barnett, V. and Roberts, D. (1993). The problem of outlier tests in survey samples. *Commun. Statist. Theor. Meth.*, **22**(10), 2703–2721. (66)

Barnett, V. and Moore, K. L. (1997). Optimal estimates in ranked-set sampling with and without perfect ordering. *J. Appl. Statist.*, **24**, 697–710. (199, 202, 203, 204)

Barnett, V. and Bown, M. (2002). Statistically meaningful standards for contaminated sites using composite sampling. *Environmetrics*, **13**, 1–13. (194, 197)

Barnett, V., Haworth, J. and Smith, T. F. M. (2002). A two phase sampling scheme with applications to auditing *or sed quis custodiet ipsos custodies. J. Roy. Statist. Soc. Series A*, **164**, 407–422. (8, 159, 160)

Barreto, M. C. M. and Barnett, V. (1999). Best linear unbiased estimators for the simple linear regression model using ranked set sampling. *J. Environ. Ecol. Statist.*, **6**(2), 119–134. (199)

Barreto, M. C. M. and Barnett, V. (2001). Estimators for a Poisson parameter using ranked set sampling. *J. Appl. Statist.*, **28**, 929–941. (199)

Bellhouse, D. R. and Rao, J. N. K. (1975). Systematic sampling in the presence of a trend. *Biometrika*, **62**, 694–697. (145)

Berg, S. (1983). Random contact processes, snowball sampling and factorial series distributions. *J. Appl. Prob.*, **20**, 31–46. (196)

Berg, S. (1988). Snowball sampling. In Kotz, S. and Johnson, N. L. (Eds) *Encyclopaedia of Statistical Sciences, Vol. 8*. Wiley, New York, pp 528–532. (193)

Biemer, P. P., Groves, R. M., Lyberg, L. E., Mathiowetz, N. A. and Sudman, S. (1991). *Measurement Errors in Surveys*. Wiley, New York. (25, 158, 161)

Biernacki, P. and Waldorf, D. (1981). Snowball: problems and techniques for chain referral sampling. *Sociology and Meth. Res.*, **10**, 141–163. (193)

Birnbaum, Z. W. and Sirken, M. G. (1965). Design of sample surveys to estimate the prevalence of rare diseases: three unbiased estimates. *Vital and Health Statistics Series 2(11)*. Government Printing Office, Washington. (192)

Boswell, M. T. and Patil, G. P. (1987). A perspective of composite sampling. *Commun. Statist. Theory Meth.*, **16**, 3069–3093. (197)

Bowan, G. L. (1994). Estimating the reduction in nonresponse bias from using a mail survey as a backup for nonrespondents to a telephone interview survey. *Res. Social Work Pract.*, **4**, 115–128. (173)

Brewer, K. R. W. and Hanif, M. (1983). *Sampling with Unequal Probabilities*. Springer Verlag, New York. (26, 54)

Bromaghin, J. and McDonald, L. (1992). Systematic encounter sampling. A simulation study. *J. Statist. Comp. Sim.*, **40**, 384. (145)

Brumelle, S., Nemetz, P. and Casey, D. (1984). Estimating means and variances of the comparative efficiency of composite and grab samples. *Environ. Monitoring Assessment*, **4**, 81–84. (197)

Buckland, S. T. (1987). On the variable circular plot method of estimating animal density. *Biometrics*, **43**, 363–384. (210)

Buckland, S. T. and Turnock, B. J. (1992). A robust line transect method. *Biometrics*, **48**, 901–909. (214)

Buckland, S. T., Anderson, D. R., Burnham, K. P. and Laake, J. L. (1993). *Distance Sampling: Estimating Abundance of Biological Populations*. Chapman and Hall, London. (26, 210, 214)

Burnham, K. P. and Anderson, D. R. (1976). Mathematical models for nonparametric inferences from line transect data. *Biometrics*, **32**, 325–336. (211, 212, 214)

Burnham, K. P., Anderson, D. R. and Laake, J. L. (1980). Estimation of density from line transect sampling of biological populations. *Wildlife Monographs 72*. Supplement to *Journal of Wildlife Management*. (210, 214)

Cane, V. R. (1966). A note on the size of epidemics and the number of people hearing a rumour. *J. Roy. Statist. Soc.*, **28**, 487–490. (193)

Carter, R. E. and Lowe, L. E. (1986). Lateral variability of forest floor properties under 2nd-growth Douglas-fir stands and the usefulness of composite sampling techniques. *Can. J. Forest Research*, **16**, 1128–1132. (194)

Cassel, C. M., Sarndal, C. E. and Wretman, J. H. (1977). *Foundations of Inference in Survey Sampling*. Wiley, New York. (25)

Chatterjee, S. (1967). A note on optimal stratification. *Skand. Akt.*, **50**, 40–44. (122)

Chaudhuri, A. and Mukerjee, R. (1987). *Randomised Response*. Marcel Dekker, New York. (178)

Chaudhuri, A. and Stenger, H. (1992). *Survey Sampling: Theory and Methods*. Marcel Dekker, New York.

Chen, S. X. (1996). Studying school effects in line transect sampling using the kerkel method. *Biometrics*, **52**, 1283–1294. (211, 214)

Chen, Y. Z. and Shao, J. (1999). Inference with survey data imputed by hot deck when imputed values are nonidentifiable. *Statistica Sinica*, **9**, 361–384. (146)

Chen, Y. Z., Rao, J. N. K. and Sitter, R. R. (2000). Efficient random imputation for missing data in complex surveys. *Statistica Sinica*, **10**, 1153–1169. (146)

Chung, J. H. and DeLury, D. B. (1950). *Confidence Limits for the Hypergeometric Distribution*. University of Toronto Press, Toronto. (56)

Ciccone, G., Forastiere, F., Agabiti, N. *et al.* (1998). Road traffic and adverse respiratory effects in children. *Occup. Environ. Med.*, **55**, 771–778. (7)

Cochran, W. G. (1977). *Sampling Techniques, 3rd Edition*. Wiley, New York. (25, 40, 44, 57, 62, 63, 74, 79, 89, 104, 105, 118, 119, 120, 121, 122, 123, 124, 125, 126, 128, 143, 145, 146, 149, 151, 161, 163, 166, 178, 184)

Cormack, R. M. (1979). Models for capture–recapture. In Cormack, R. M., Patil, G. P. and Robson, D. S. (Eds). *Sampling Biological Populations*. International Co-operative Publishing House, Fairland, Maryland, pp 217–255. (217)

Cormack, R. M., Patil, G. P. and Robson, D. S. (Eds) (1979). *Sampling Biological Populations*. International Co-operative Publishing House, Fairfield, Maryland. (26)

Couper, M. P., Baker, R. P., Bethlehem, J., Clark, C. Z. F., Martin, J., Nicholis, W. L. and O'Reilly, J. M. (1998). *Computer Assisted Survey Information Collection*. Wiley, New York. (172)

Cox, B. G. and Cohen, S. B. (1985). *Methodological Issues for Health Care Surveys*. Marcel Dekker, New York.

Cox, B. G., Binder, D. A., Chinnappa, B. N., Christianson, A., Colledge, M. J. and Kott, P. S. (1995). *Business Survey Methods*. Wiley, New York. (26)

Cox, D. R. (1952). Estimation by double sampling. *Biometrika*, **39**, 217–227. (44)

Cox, D. R. and Snell, E. J. (1979). On sampling and the estimation of rare errors. *Biometrika*, **66**, 125–132. (191)

Cunliffe, S. Y. and Goldstein, H. (1979). Ethical aspects of survey research. *Appl. Statist.*, **28**, 219–222. (166)

Curtis, J. and Sparrow, N. (1997). How accurate are traditional quota opinion polls? *J. Marketing Research*, **39**, 433–448. (126)

Dalenius, T. (1957). *Sampling in Sweden*. Contributions to the methods and theories of sample survey practice. Almqvist and Wicksell, Stockholm. (125)

Daley, D. J. and Kendall, D. G. (1965). Stochastic rumours. *J. Inst. Math. Appl.*, **19**, 42–55. (193)

David, H. A. (1981). *Order Statistics, 2nd Edition*. Wiley, New York. (202)

Davis, A. C., Lovell, E. A., Smith, P. A. and Ferguson, M. A. (1998). The contribution of social noise to tinnitus in young people – a preliminary report. *Noise and Health*, **1**, 40–46. (9)

Dell, T. R. and Clutter, J. L. (1972). Ranked set sampling theory with order statistics background. *Biometrics*, **28**, 545–55. (199, 200, 204)

Deming, W. E. (1990). *Sampling Design in Business Research*. Wiley, New York. (26)

Dillman, D. A. (1976). *Mail and Telephone Surveys; the Total Design Method*, Wiley, New York. (171, 173)

Dillman, D. A. (2000). *Mail and Internet Surveys; the Tailored Design Method, 2nd Edition*. Wiley, New York. (25, 171, 172, 173)

Dorfman, R. (1943). Detection of defective members of a large population. *Ann. Math. Statist.*, **14**, 436–488. (194)

Dowling, T. A. and Schactman, R. H. (1975). On the relative efficiency of randomised response models. *J. Amer. Statist. Assoc.*, **70**, 84–87. (179)

Dunn, R. and Harrison, A. R. (1993). Two-dimensional systematic sampling of land use. *J. Roy. Statist. Soc. Series C*, **42**, 585–601. (145)

Eberhardt, L. L. (1978). Transect methods for population studies. *J. Wildlife Management*, **42**, 1–31. (210, 214)

Elder, R. S., Thompson, W. O. and Myers, R. H. (1980). Properties of composite sampling procedure. *Technometrics*, **22**, 179–186. (197)

Erdos, P. L. (1970). *Professional Mail Surveys*. McGraw-Hill, New York. (172)

Ericksen, E. P. (1976). Sampling a rare population: a case study. *J. Amer. Statist. Assoc.*, **71**, 816–822. (192)

Fisher, R. A. (1934). The effects of methods of ascertainment upon the estimation of frequencies. *Ann. Eugenics.*, **6**, 13–25. (208)

Fisher, R. A. and Yates, F. (1973). *Statistical Tables for Biological, Agricultural and Medical Research, 6th Edition*. Longman, Harlow. (57)

Ford, B. M. (1983). An overview of hot-deck procedures. In Madow, W. G., Olkin, I. and Rubin, D. B. (Eds) *Incomplete Data in Sample Surveys, Vol. 2*. Academic Press, New York, pp 185–207. (185)

Fowler, F. J. and Mangione, T. W. (1990). *Standardized Survey Interviewing: Minimizing Interviewer-related Error*. Sage, Newbury Park, California. (171)

Freund, J. E. (1992). *Mathematical Statistics, 5th Edition*. Prentice Hall, Eaglewood Cliffs, NJ. (4)

Gani, J. and Jerwood, D. (1971). Markov chain methods in chain binomial epidemic models. *Biometrics*, **57**, 591–603. (193)

Garner, F. C., Stapanian, M. A. and Williams, L. R. (1988). Composite sampling for environmental monitoring. In Keith, L. H. (Ed.) (1986) *Principles of Environmental Sampling*. American Chemical Society, Washington, pp 363–374. (194)

Gates, C. E. (1979). Line transect and related issues. In Cormack, R. M., Patil, G. P. and Robson, D. S. (Eds) *Sampling Biological Populations*. International Co-operative Publishing House, Burtonsville, Maryland, pp 71–154. (214)

Gilbert, R. O. (1987). *Statistical Methods for Environmental Pollution Monitoring*. Van Nostrand Reinholt, New York. (26)

Goodman, L. A. (1961). Snowball sampling. *Ann. Math. Statist.*, **32**, 148–170. (193)

Green, E. J. and Strawderman, W. E. (1986). Reducing sample size through the use of a composite estimator – an application to timber volume estimation. *Can. J. Forest Research*, **16**, 1116–1118. (194)

Greenberg, B. G. *et al.* (1969). The unrelated question randomised response model: theoretical model. *J. Amer. Statist. Assoc.*, **64**, 520–539. (178)

Greenberg, B. G. *et al.* (1971). Application of the randomised response technique in obtaining quantitative data. *J. Amer. Statist. Assoc.*, **66**, 243–250. (179)

Groves, R. M. (1989). *Survey Errors and Survey Costs*. Wiley, New York. (25, 158, 161, 163, 164, 171, 177, 181)

Groves, R. M. and Couper, M. P. (1998). *Nonresponse in Household Interview Surveys*. Wiley, New York. (25)

Groves, R. M., Biemer, P. P., Lyberg, L. E., Massey, J. T., Nicholls, II W. L. and Waksberg, J. *et al.* (Eds) (1988). *Telephone Survey Methodology*. Wiley, New York. (25, 171, 173, 188)

Groves, R. M., Dillman, D., Eltinge, J., and Little, R. J. A. (2001). *Survey Nonresponse*, Wiley, New York. (173)

Hajek, J. (1981). *Sampling from a Finite Population*. Marcel Dekker, New York. (25, 158)

Hannan, E. J. (1962). Systematic sampling. *Biometrika*, **49**, 281–283. (145)

Hansen, M. H. and Hurwitz, W. N. (1943). On the theory of sampling from finite populations. *Ann. Math. Statist.*, **14**, 333–362. (53)

Hansen, M. H. and Hurwitz, W. N. (1946). The problem of nonresponse in sample surveys. *J. Amer. Statist. Assoc.*, **41**, 517–529. (185)

Hansen, M. H., Hurwitz, W. N. and Madow, W. G. (1993). *Sample Survey Methods and Theory, Vols 1 and 2*. Wiley, New York. (25)

Hartley, H. O. (1966). Systematic sampling with unequal probabilities and without replacement. *J. Amer. Statist. Assoc.*, **61**, 739–748. (145)

Hayne, D. W. (1949). An examination of the strip census method for estimating animal populations. *J. Wildlife Management*, **13**, 145–157. (210)

Hedayat, A. S. and Sinha, B. K. (1991). *Design and Inference in Finite Population Sampling*. Wiley, New York. (25, 44, 53, 54, 62, 91, 121, 124, 166, 171, 184)

Hedges, B. M. (1979). Sampling minority populations. In Wilson, M. J. (Ed.) *Social and Educational Research in Action*. Longman, London. (192)

Heilbron, D. C. (1978). Comparison of estimators of the variance of systematic sampling. *Biometrika*, **65**, 429–433. (145)

Hoinville, G., Jowell, R. *et al.* (1978). *Survey Research Practice*. Heinemann, London. (25, 26, 158, 164, 171, 172, 177)

Horvitz, D. G. and Thompson, D. J. (1952). A generalisation of sampling without replacement from a finite universe. *J. Amer. Statist. Assoc.*, **47**, 663–685. (53)

Horwitz, D. G., Shah, B. V. and Simmons, W. R. (1967). The unrelated random response model. *Proc. Soc. Statist. Sec. Amer. Statist. Assoc.*, pp 65–72. (178)

Huff, D. and Geis, I. (1993). *How to Lie with Statistics*. Norton, New York. (4)

Huntsberger, D. V. and Billingsley, P. (1987). *Elements of Statistical Inference*. Allyn and Bacon, Boston. (4)

Jessen, R. J. (1978). *Statistical Survey Techniques*. Wiley, New York. (25)

Journel, A. G. (1988). Non-parametric geostatistics for risk and additional sampling assessment. In Kieth, L. (Ed.) *Principles of Environmental Sampling*. American Chemical Society, pp 45–72. (26)

Kalton, G. (1977). Practical methods for estimating survey sampling errors. *Bull. Int. Statist. Inst.*, **47**(3), 495–514. (182, 183)

Kalton, G. (1983). *Compensating for Missing Survey Data*. Research Report Series, Institute for Social Research, University of Michigan. (25)

Kalton, G. and Anderson, D. W. (1986). Sampling rare populations. *J. Roy Statist. Soc. A*, **149**, 65–82. (191, 192)

Kalton, G. and Kasprzyk, D. (1986). The treatment of missing survey data. *Survey Methodology*, **12**, 1–16. (185)

Kalton, G. Collins, M. and Brook, L. (1978). Experiments in wording opinion questions. *Applied Statistics*, **27**, 149–161. (176)

Kaplan, C. D., Korf, D. and Sterk, C. (1987). Temporal and social contexts of heroin-using populations – an illustration of the snowball sampling technique. *J. Nervous and Mental Disease*, **175**, 566–574. (193)

Kasprzyk, D., Duncan, G, Kalton, G. and Singh, M. P. (1989). *Telephone Survey Methodology*. Wiley, New York. (174)

Kennedy, J. M., Kuh, G. D. and Carini, R. (2000). Web and mail surveys: preliminary results of comparisons based on a large-scale project. *Report, Annual Meeting of the American Association for Public Opinion Research*. Portland, Oregon. (172)

Kish, L. (1995). *Survey Sampling*. Wiley, New York. (25)

Kish, L. and Frankel, M. (1974). Inference from complex samples (with Discussion). *J. Roy. Statist. Soc., Series B*, **36**, 1–37. (146, 149)

Kittleson, M. J. (1995). An assessment of the respose rate via the postal service and e-mail. *J. Health Behavior, Education and Promotion*, **19**(2), 27–29. (172)

Korn, E. L. and Graubard, B. I. (1999). *Analysis of Health Surveys*. Wiley, New York. (26)

Lahiri, D. B. (1951). A method for sample selection providing unbiased ratio estimates. *Bull. Int. Statist. Inst.*, **33**(2) 133–140. (95)

Lancaster, V. A. and McNulty, S. (1998). A review of composite sampling methods. *J. Amer. Statist. Assoc.*, **93**, 1216–1230. (197)

Lee, E. L., Forthofer, R. N. and Lorimer, R. J. (1989). *Analyzing Complex Survey Data*. Sage, Beverly Hills, CA. (146, 149, 183)

Lehtonen, R. and Pahkinen, E. J. (1995). *Practical Methods for Design and Analysis of Complex Surveys*. Wiley, Chichester. (26, 134, 146, 149, 151, 182, 183, 185)

Lepkowski, J. M. (1991). Sampling the difficult-to-sample. *Amer. J. Nutrition*, **121**, 416–423. (192)

Lepkowski, J. and Bowles, J. (1996). Sampling error software for personal computers. *Survey Statistician, No. 35*, 10–17. (146)

Lessler, J. T. and Kalsbeek, W. D. (1992). *Nonsampling Error in Surveys*. Wiley, New York. (25, 158, 161, 163, 177, 185)

Levy, P. S. (1977). Optimum allocation in stratified random network sampling for estimating the prevalence of attributes in rare populations. *J. Amer. Statist. Assoc.*, **72**, 758–763. (192)

Levy, P. S. and Lemeshow, S. (1980). *Sampling for Health Professionals*. Lifetime Learning, Belmont, CA. (26)

Levy, P. S. and Lemeshow, S. (1991). *Sampling of Populations: Methods and Applications, 3rd Edition*. Wiley, New York. (25, 121, 124, 128, 134, 143, 146, 164, 171, 172, 177, 181, 184, 185)

Lloyd, E. H. (1952). Least-squares estimation of location and scale parameters using order statistics. *Biometrika*, **34**, 41–67. (202)

Lovison, G., Gore, S. D. and Patil, G. P. (1994). Design and analysis of composite sampling procedures: a review. In Patil, G. P. and Rao, C. R. (Eds) *Handbook of Statistics, Vol. 12*. Elsevier, Amsterdam, pp 103–166. (194)

Loynes, R. M. (1976). Asymptotically optimal randomised response procedures. *J. Amer. Statist. Assoc.*, **71**, 924–928. (178)

Luntz, F. I. (1994). Voices of Victory, Part I: focus group research in American politics. Part II: the makings of a good focus group. *The Polling Report,* May 16 and May 30, 1994. (175)

Lyberg, L. E., Biemer, P., Collins, M., De leeuw, E., Dippo, C., Schwarz, N. and Trewin, D. (1997). *Survey Measurement and Process Quality.* Wiley, New York. (158, 161, 163)

Lynn, P. and Jowell, R. (1996). How might opinion polls be improved? The case for probability sampling. *J. Roy. Statist. Soc., Series A*, **159**, 21–39. (126)

Mangione, T. W. (1995). *Mail Surveys: Improving the Quality.* Sage, Thousand Oaks, CA. (171)

Manly, B. F. J. and McDonald, L. L. (1996). Sampling wildlife populations, *Chance*, **9**(2), 9–20. (214, 217)

Mason, R. and Traugott, M. W. (2000). Impact of item nonresponse on nonsampling error. *Paper presented to the International conference on Survey Nonresponse, Portland, Oregon 28–31 October 1999*, www.jpsm.umd.edu/icsn.papers/mason.htm. (174, 186)

Mavis, B. E. and Brocato, J. J. (1998). Postal surveys versus electronic mail surveys – the tortoise and the hare revisited. *Evaluation and the Health Professions*, **21**(3), 395–408. (172)

Maynard, D. W., Houtkoop-Steenstra, H., Schaeffer, N. C. and Van der Zouwen, J. (2001). *Standardization and Tacit Knowledge.* Wiley, New York.

McIntyre, G. A. (1952). A method of unbiased selective sampling, using ranked sets. *Aust. J. Agric. Res*, **3**, 88–95. (198, 199)

Mohlin, B., Pilley, J. R. and Shaw, W. C. (1991). A survey of craniomandibular disorders in 1000 12-year-olds – study design and base-line data in a follow-up-study. *European J. Orthodontics*, **13**, 111–123. (192)

Mood, A. M., Graybill, F. A. and Boes, D. C. (1974). *Introduction to the Theory of Statistics, 3rd Edition.* McGraw-Hill, Kogakusha. (4)

Moore, D. S. (2001). *Statistics: Concepts and Controversies, 5th Edition.* Freeman, New York. (4)

Moore, D. S. and McCabe, G. P. (1999). *Introduction to the Practice of Statistics, 3rd Edition.* Freeman, New York. (4)

Moser, C. A. and Kalton, G. (1971/1999). *Survey Methods in Social Investigation, 2nd Edition.* Heinemann, London. (25, 26, 158, 164, 171, 172, 181)

Nathan, G. (1976). An empirical study of response and sampling errors for multiplicity estimates with different counting rules. *J. Amer. Statist. Assoc.*, **71**, 808–815. (192)

Neave, H. R. (1978). *Statistical Tables.* George Allen and Unwin, London. (56)

Neyman, J. (1934). On the two different aspects of the representative method: the method of stratified sampling and the method of purposive selection. *J. Roy. Statist. Soc.*, **97**, 558–606. (112)

Neyman, J. (1938). Contribution to the theory of sampling human populations. *J. Amer. Statist. Assoc.*, **33**, 101–116. (119)

Nunnally, J. C. (1967). *Psychometric Theory*. McGraw-Hill, New York. (26)

O'Neil, M. J. (1979). Estimating the nonresponse bias due to refusals in telephone surveys. *Public Opinion Quarterly*, **143**, 219–232. (173)

Owen, D. B. (1962). *Handbook of Statistical Tables*. Addison-Wesley, Reading, Mass. (56)

Patil, G. P. (1991). Encountered data, statistical ecology, environmental statistics and weighted distribution methods. *Environmetrics*, **2**, 377–433. (168, 190, 191)

Patil, G. P. (1997). Weighted distributions. In Armitage, P. and Colton, T. (Eds) *Encyclopaedia of Biostatistics*. Wiley, New York. (208)

Patil, G. P. and Rao, C. R. (1977). Weighted distributions and a survey of their applications. In Krishnaiah, P. R. (Ed.) *Applications of Statistics*. North-Holland, Amsterdam. (209)

Patil, G. P., Rao, C. R. and Zelen, M. (1988). Weighted distributions. In Kotz, S. and Johnson, N. L. (Eds) *Encyclopaedia of Statistical Sciences, Vol. 9*. Wiley, New York. (208)

Patil, G. P., Gore, S. D. and Sinha, A. H. (1992). Environmental sampling and statistical modelling with examples. *Proc. XVIth International Biometric Conference, Hamilton, Ontario*. (194)

Peterson, R. A. (2000). *Constructing Effective Questionnaires*. Sage, Thousand Oaks, CA. (171)

Pfefferman, D., Skinner, C. J., Holmes, D. J. Goldstein, H. and Rasbash, J. (1998). Weighting for unequal selection probabilities in multilevel models. *J. Roy. Statist. Soc. Series B*, **60**, 23–40. (146)

PNMD (1989). Statistical models and analyses in auditing. *Statistical Science*, **4**(1), 2–23. (8)

Pollock, K. H. (1978). A family of density estimators for line transect sampling. *Biometrics*, **34**, 475–487. (214)

Pollock, K. H. (1991). Modelling capture, recapture and removal statistics for estimation of demographic parameters for fish and wildlife populations: past, present and future. *J. Amer. Statist. Assoc.*, **86**, 225–238. (217)

Quang, P. X. (1991). A nonparametric approach to size-biased line transect sampling. *Biometrics*, **47**, 269–279. (210)

Quang, P. X. (1993). Nonparametric estimators for variable circular plot surveys. *Biometrics*, **49**, 837–852.

Raj, D. (1972). *Design of Sample Surveys*. McGraw-Hill, New York. (25)

Ramsey, F. L. and Scott, J. M. (1979). Estimating population densities from variable circular plot surveys. In Cormack, R. M., Patil, G. P. and Robson, D. S. (Eds) *Sampling Biological Populations*. International Co-operative Publishing House, Burtonsville, Maryland. (210)

Ramsey, F., Gates, C., Patil, G. P. and Taillie, C. (1988). On transect sampling to assess wildlife population and marine resources. In Krishnaiah, P. R. and Rao, C. R. (Eds) *Handbook of Statistics, Vol. 6*. Elsevier, B.V., pp 515–532. (210)

Rao, C. R. (1965). On discrete distributions arising out of methods of ascertainment. In Patel G. P. (Ed.) *Classical and Contagious Discrete Distributions*, Pergamon Press, Calcutta, pp 320–333. (208, 209)

Rao, J. N. K. (1996). On variance estimation with imputed sample data. *J. Amer. Statis. Assoc.*, **91**, 499–506. (146)

Rao, J. N. K. (1999). Some current trends in sample survey theory and methods. *Sankhya, Series B*, **61**, 1–57. (66)

Rao, P. S. R. S. (1988). Ratio and regression estimators. In Krishnaiah, P. R. and Rao, C. R. (Eds) *Handbook of Statistics, Vol. 6 (Sampling)*. Elsevier, Amsterdam, pp 449–468. (79, 89)

Ridout, M. S. and Cobby, J. M. (1987). Ranked set sampling with non-random selection of sets and errors in ranking. *Applied Statistics*, **36**, 145–152. (199, 200)

Robinson, J. (1987). Conditioning ratio estimates under simple random sampling. *J. Amer. Statist. Assoc.*, **82**, 826–831. (79)

Rohde, C. A. (1976). Composite sampling. *Biometrics*, **32**, 273–282. (197)

Rubin, D. B. (1987). *Multiple Imputation for Nonresponse in Surveys*. Wiley, New York. (185, 186)

Sampford, M. R. (1962). *An Introduction to Sampling Theory with Applications to Agriculture*. Oliver and Boyd, Edinburgh. (26, 118)

Särndal, D.-E., Swensson, B. and Wretman, J. (1992). *Model Assisted Survey Sampling*. Springer Verlag, New York. (66)

Schaeffer, R. L., Mendenhall, W. and Ott, L. (1986). *Elementary Survey Sampling, 3rd Edition*. Duxbury, Boston. (25)

Schuman, H. and Presser, S. (1981). *Questions and Answers in Attitude Surveys: experiments on question form, wording and context*. Academic Press, New York. (177)

Scott, A. and Wild, C. (2001). Case-control studies in complex sampling. *J. Roy. Statist. Soc. Series C*, **50**, 389–401. (146)

Seber, G. A. F. (1982). *The Estimation of Animal Abundance and Related Parameters, 2nd Edition*. Griffin, London. (26, 210, 214, 216)

Seber, G. A. F. (1986). A review of estimating animal abundance. *Biometrics*, **42**, 267–292. (26, 210, 215, 217)

Seber, G. A. F. (1992). A review of estimating animal abundance II. *International Statistical Review*, **60**, 129–166. (210, 217)

Shao, J. and Sitter, R. R. (1996). Bootstrap for imputed survey data. *J. Amer. Statist. Assoc.*, **91**, 1278–1288. (146)

Shao, J. and Chen, Y. Z. (1998). Bootstrapping sample quantiles based on complex survey data under hot deck imputation. *Statistica Sinica*, **8**, 1071–1085. (146)

Shao, J., Chen, Y. and Chen, Y. Z. (1998). Balanced repeated replication for stratified multistage survey data under imputation. *J. Amer. Statist. Assoc.*, **93**, 819–831. (147)

Singh, D. and Chaudhury, F. S. (1986). *Theory and Analysis of Sample Survey Designs*. Wiley Eastern, New Delhi. (25, 124)

Sinha, B. K., Sinha, B. K. and Purkayastha, S. (1996). On some aspects of ranked set sampling for estimation of normal and exponential parameters. *Statist. Decisions*, **14**, 223–240. (99, 202, 203)

Sirken, M. G. (1970). Household surveys with multiplicity. *J. Amer. Statist. Assoc.*, **65**, 257–266.

Sirken, M. G., Grubard, B. J. and La Valley, R. W. (1980). Evaluation of census population coverage by network surveys. *Proc. Survey Res. Meth. Sect., Amer. Statist. Assoc.*, pp 239–244. (192)

Skinner, C. J., Holt, D. and Smith, T. M. F. (Eds) (1989). *Analysis of Complex Surveys.* Wiley, New York. (26, 134, 146, 149, 151, 182)

Smith, T. M. F. (1976). *Statistical Sampling for Accountants.* Accountancy Age, Haymarket Publishing, London. (8, 26)

Smith, T. M. F. (1994). Sample surveys 1975–1990; an age of reconciliation. *Int. Statist. Rev.*, **62**, 3–34. (66)

Smith, T. M. F. (2000). Some recent developments in sample survey theory and their relevance to official statisticians. *Proc. Int. Assoc. Survey Statisticians.* 52nd ISI session, Helsinki, pp 3–15. (8, 66)

Steele, A. (1992). *Audit Risk and Audit Evidence; The Bayesian Approach to Statistical Auditing.* Academic Press, New York. (8)

Stehman, S. V. and Overton, W. S. (1994). Environmental sampling and monitoring. In Patil, G. P. and Rao, C. R. (Eds) *Handbook of Statistics, Vol. 12.* Elsevier, Amsterdam, pp 263–306. (26)

Stephan, F. and McCarthy, P. J. (1958). *Sampling Opinions.* Wiley, New York. (126)

Stokes, S. L. (1977). Ranked set sampling with concomitant variables. *Commun. Statist. Theor. Meth.*, **6**, 1207–1211. (199, 204)

Stokes, S. L. (1980). Estimation of variance using judgement ordered ranked-set samples. *Biometrics*, **36**, 35–42. (199, 204)

Stokes, S. L. (1995). Parametric ranked set sampling. *Ann. Inst. Statist. Math.*, **47**, 465–482. (199)

Stuart, A. (1984). *The Ideas of Sampling, Revised edition.* Griffin, London. (25)

Sudman, S. (1972). On sampling very rare human populations. *J. Amer. Statist. Assoc.*, **67**, 335–339. (191, 192)

Sudman, S. (1976). *Applied Sampling.* Academic Press, New York. (158)

Sudman, S. (1996). Probability sampling with quotas. *J. Amer. Statist. Assoc.*, **61**, 749–771. (126)

Sudman, S., Sirken, M. G. and Cowan, C. D. (1988). Sampling rare and elusive populations. *Science*, **240**, 991–996. (192)

Takahasi, K. and Wakimoto, K. (1968). On unbiased estimates of the population mean based on the sample stratified by means of ordering. *Ann. Inst. Statist. Math.*, **20**, 1–31. (199)

Thompson, S. K. (1992). *Sampling.* Wiley, New York. (25, 40, 53, 54, 57, 119, 120, 121, 124, 143, 146, 149, 183, 184, 190, 210, 217)

Thompson, S. K. and Seber, G. A. F. (1996). *Adaptive Sampling.* Wiley, New York. (206)

Tschuprow, A. A. (1923). On the mathematical expectation of the moments of frequency distributions in the case of correlated observations. *Metron*, **2**, 461–493, 646–683. (112)

Tse, A. C. B. (1998). Comparing the response rate, response speed and response quality of two methods of sending questionnaires: e-mail vs. mail. *J. Market Res. Soc.*, **40**(4), 353–361. (172)

Tse, A. C. B. *et al.* (1995). Comparing two methods of sending out questionnaires: e-mail versus mail. *J. Market Res. Soc.*, **37**(4), 441–445. (172)

Useem, M. (1973). *Conscription Protest and Social Conflict.* Wiley, New York. (193)

Valliant, R. (1990). Comparisons of variance estimators in stratified sampling and in systematic sampling. *J. Official Statist.*, **6**(2), 115–131. (145)

Valliant, R., Dorfman, A. H. and Royall, R. M. (2000). *Finite Population Sampling and Inference: A Predictive Approach.* Wiley, New York. (66)

Warner, S. L. (1965). Randomised response: a survey technique for eliminating evasive answer bias. *J. Amer. Statist. Assoc.*, **60**, 63–69. (178)

Warner, S. L. (1971). The linear randomised response model. *J. Amer. Statist. Assoc.*, **66**, 884–888. (179)

Watson, G. H. (1936). A study of the group screening method. *Technometrics*, **3**, 371–388. (194)

Webster, R. and Oliver, M. (2000). *Geostatistics for Environmental Scientists.* Wiley, Chichester. (26)

Wetherill, G. B. (1972). *Elementary Statistical Methods, 2nd Edition.* Chapman and Hall, London. (4)

Whitemore, A. S. (1997). Multistage sampling designs and estimating equations. *J. Roy. Statist. Soc. Series B*, **59**, 589–602. (146)

Wolter, K. M. (1985). *Introduction to Variance Estimation.* Springer Verlag, New York. (26, 182, 183)

Yates, F. (1981). *Sampling Methods for Censuses and Surveys, 4th Edition.* Griffin, London. (25)

Yung, W. and Rao, J. N. K. (2000). Jackknife variance estimation under imputation for estimators using poststratification information. *J. Amer. Statist. Assoc.*, **95**, 903–915. (146)

Zinger, A. (1980). Variance estimation in partially systematic sampling. J. Amer. Statist. Assoc., **75**, 206–211. (145)

Index

Note: Where one heading provides greater detail, cross-references are given from other headings. Where subjects are discussed over several consecutive pages, a page-range is given. Page numbers in **bold type** refer to major development of the subject.